Solid Mechanics and Its Applications

Volume 248

Aims and Scope of the Series

The fundamental questions arising in mechanics are: *Why*?, *How*?, and *How much*? The aim of this series is to provide lucid accounts written by authoritative researchers giving vision and insight in answering these questions on the subject of mechanics as it relates to solids.

The scope of the series covers the entire spectrum of solid mechanics. Thus it includes the foundation of mechanics; variational formulations; computational mechanics; statics, kinematics and dynamics of rigid and elastic bodies: vibrations of solids and structures; dynamical systems and chaos; the theories of elasticity, plasticity and viscoelasticity; composite materials; rods, beams, shells and membranes; structural control and stability; soils, rocks and geomechanics; fracture; tribology; experimental mechanics; biomechanics and machine design.

The median level of presentation is to the first year graduate student. Some texts are monographs defining the current state of the field; others are accessible to final year undergraduates; but essentially the emphasis is on readability and clarity.

More information about this series at http://www.springer.com/series/6557

Jaap Wijker

Miles' Equation in Random Vibrations

Theory and Applications in Spacecraft Structures Design

Springer

Jaap Wijker
Department of Applied Mechanics
University of Twente
Velserbroek, Noord-Holland
The Netherlands

ISSN 0925-0042 ISSN 2214-7764 (electronic)
Solid Mechanics and Its Applications
ISBN 978-3-319-73113-1 ISBN 978-3-319-73114-8 (eBook)
https://doi.org/10.1007/978-3-319-73114-8

Library of Congress Control Number: 2017963000

Printed on acid-free paper

This Springer imprint is published by Springer Nature
The registered company is Springer International Publishing AG
The registered company address is: Gewerbestrasse 11, 6330 Cham, Switzerland

This book is dedicated to Jac. Zuurbier (1942–2016)

Foreword

Let me start this foreword by introducing myself. I got my Ph.D. in Mechanical Engineering at the Noise and Vibration Research Group at KU Leuven, Belgium, in 1990. After that, I decided to move to industry. The research group that I worked in had been in contact with Fokker Space and Systems in Amsterdam-Zuidoost, and I applied for a job with them in their structural engineering department. I have created my finite element models, I have run many finite element problems, and I have developed tools for preprocessing finite element models and tools for post-processing finite element results. I have worked mainly for the development of the Ariane 5 engine frame, for solar panels on several spacecraft, and I have also spent some time at Fokker Aircraft for the development of the F70.

With the background that I had built up in doing research at KU Leuven, the plain logic of the HR department at Fokker Space was to employ me in a small special branch of their engineering department, which was in charge of non-routine technical analysis runs on development projects. At that time, the other members of the team were Marcel Ellenbroek en Simon Appel, who were much more clever than me and who were very proficient at getting much more information out of a finite element run than the f06 file provided at first sight. That understanding of technical problems and design features was very helpful in the development of advanced technical products for the space sector. Although we did not work directly for customers, we were probably the most effective team of the company.

The leader of this gang was Jaap Wijker. He favored technical competence and thorough understanding of physical phenomena in the design of structures over hierarchy and other managerial blabla. His instructions on the way to handle our tasks were clear and explicit on the technical details of element selection, matrix manipulations, definition of boundary conditions, etc. He was the leader by example, not the manager who monitored the key performance indicators of his staff. I am not sure that the KPI word was common in those days, but Jaap was definitely not the man who invented it.

After two years of full-time employment at Fokker Space and Systems, I switched to a part-time regime. In the other part, I rejoined KU Leuven. Work in Jaap's group was a perfect extension of academic work, and that year has been the best period of my professional lifetime, when working in both an industrial environment and in academia. I will always be grateful to Jaap for giving me that opportunity. Unfortunately, the practical organization of spending three days in Amsterdam and two days in Leuven was rather impractical and after one year I went full time at KU Leuven. Jaap followed a similar trajectory, while maintaining his position at Fokker Space, he started teaching in a part-time role at TU Delft.

While I am the engineer who is employed at university, Jaap was the academic who worked in industry.

And Jaap fully deployed his academic interest only after his retirement from Dutch Space, when he started Ph.D. research under the supervision of André de Boer at TU Twente and with the guidance of Marcel, his former team member at Fokker. It is an amazing achievement to complete Ph.D. studies at the age of 72 … or no, it is not that amazing … There is a very clear logic. I now realize that he started his Ph.D. on his first working day at Fokker! Jaap has lived structural dynamics all his life, and he will continue to do so for the years to come. Every spacecraft is nothing else but an assembly of mass, spring and damper elements, discrete or continuous, which is excited by pure or swept sine, by white or colored noise, with spectra and notches. He has put all his passion in writing this book which is the compilation of 45 years of experience.

Enjoy reading!

Leuven, Belgium
August 2017

Dirk Vandepitte

Preface

After my doctorate at the University of Twente, the Netherlands, on October 19, 2017, my friends and family expected from me to fall into a black hole, but instead of that the idea grows to write a book about Miles' equation and applications in spacecraft structures design. In my active working live at Fokker Space and Systems BV., Dutch Space BV nowadays called Airbus Defence and Space (ADS), the Miles' equation was applied frequently to estimate the r.m.s. acceleration response of a spacecraft, instrument, box, etc., when exposed to random vibrations.

Velserbroek, The Netherlands — Jaap Wijker
April 2017

Acknowledgements

I stayed many times too long in my study room or even during holidays writing and checking the manuscript about Miles' equation. Of course, my wife Wil complained but she showed also her understanding about my compassion to write this book, and therefor I gratefully thank Wil.

Contents

Acronyms

In this chapter, a list of abbreviations and symbols is provided.

A, A_o, A_p	(Surface) area (m^2)
$[A_o]$	State matrix
ASD	Acceleration spectral density
(B)	Filter vector
a, b	length (radius) and width panel, circular plate (m)
c	Damper constant of SDOF system (Ns/m), constant Wöhler (s-N) curve, weighting factor
$[C]$	Damping matrix
d	(m) diameter
dB	Decibel
$D, \bar{D}, D_i$	Bending rigidity of plate or shell (Nm^2/m), shear force (N, N/m), damage ratio, diameter, Dirlik parameters
DK	Dirlik method
E	Young's modulus (Pa)
E(-)	Expected value
ERS	Extreme response spectrum
f	Frequency (Hz)
f_c, f_{low}, f_{up}	Center frequency, lower and upper frequency (Hz)
f_n	Natural frequency (Hz)
F	Force (N)
FDS	Fatigue damage spectrum
FEA	Finite element analysis
FEM	Finite element model
g	Gravitational acceleration (1g = 9.81 m/s^2), structural (hysteric) damping factor
h	Sandwich core height (m)
$H(\omega), H(f)$	Frequency response function (FRF)
i, j	$\sqrt{-1}$

I	Second moment of area (m^4)
k	Spring stiffness of SDOF system (N/m)
K_s	Shape factor
$[K]$	Stiffness matrix
L	Length of bar or beam (m)
L/V	Launch Vehicle
m	Mass of SDOF system (kg), mass per unit of length (kg/m), mass per unit of area plate or shell (kg/m^2), meter
m, n	Wave numbers
m_i	Spectral moment
M, M_p	Discrete mass, total panel mass (kg), bending moment (Nm, Nm/m)
M_{eff}	Modal effective mass (kg)
$[M]$	Mass matrix
n_i	Number of cycles
N_i	Number of allowable cycles in s-N curve
NB	Narrow band method
Pa	Pascal (N/m^2)
p_{ref}	Reference pressure
p	Pressure (Pa), joint probability density function
P	Random pressure (Pa)
$[P]$	Correlation matrix
PSD	Power spectral density
PCB	Printed circuit board
Q	Amplification, quality factor (-), Dirlik parameter
Q_s	First moment of area
r	Radius (m)
rms	Root-mean-square (index)
r.m.s.	Root-mean-square
$R(\tau)$	Correlation function
RRS	Random response spectrum
s	Stress s-N curve
SB	Steinberg method
S/C	Spacecraft
SDOF	Single degree of freedom
SPL	Sound pressure level (dB)
SRS	Shock response spectrum
t	Plate (shell) thickness
T	Total time (s), kinetic energy
u	Displacement
(u)	Assumed mode in Rayleigh quotient
$\dot{u}$	Velocity
$\ddot{U}$	Base acceleration (m/s^2)
U	Strain energy
VRS	Vibration response spectrum

w, W	Deflection (m)
W_a	Power spectral density of acceleration response (g^2/Hz)
W_F	Power spectral density of force (N^2/Hz)
W_p	Power spectral density of pressure field (Pa^2/Hz)
$W_{\ddot{U}}$	Power spectral density of base acceleration (g^2/Hz, $(m/s^2)^2$/Hz))
x	(1/3) Octave ratio
x_{stat}	Static deformation
X	Random displacement (m)
$\dot{X}$	Random velocity (m/s)
$\ddot{X}$	Random acceleration (m/s^2)
Y	State variable
z	z-plane variable
Z	Random relative displacement (m), normalized amplitude
$\dot{Z}$	Random relative velocity (m/s)
$\ddot{Z}$	Random relative acceleration (m/s^2)
α	Parameter, factor
β	Parameter, factor
$\ddot{\gamma}$	Load factor (g)
Γ	Gamma function, mode (modal) participation factor
ε	Strain tensor, fraction
δ	Displacement (m), Kronecker delta function, spectral parameter, Dirac delta function
Δ	Difference
ζ	Modal damping ratio (-)
η	Generalized coordinate
ν, ν_o, ν_p	Poisson's ratio (-), expected frequency, expected peak frequency (Hz)
π	Pi ($\pi = 3.14159\cdots$)
ρ	Density material (kg/m^3)
σ	Stress tensor, standard deviation
τ	Time shift (s), shear stress (Pa)
ϕ	Vibration mode, mode shape
Φ	Normal cumulative distribution, diameter potting
ψ	Spatial distribution function
ω	Radial frequency (rad/s)
ω_n	Radial natural frequency (rad/s)

Chapter 1
Introduction

Abstract A general introduction about the book 'Miles Equation in Random Vibrations' is given, and the contents of chapters and appendices are summarized.

In my opinion, Miles' equation should be part of the toolbox of every mechanical engineer working in spacecraft structures design and analysis. To recognize the limitation(s) in applying Miles' equation the engineer should be required to derive this equation when using it for the first time: applying either the spectral analysis approach or Lyapunov's equation or applying the Fokker–Planck–Kolmogorov equation (FPKE).

In 1954, John W. Miles published in the Journal of the Aeronautical Sciences his famous article entitled: 'On structural fatigue under random loading' [1]. In this article, he derived the root-mean-square (r.m.s.) response of a single degree of freedom (SDOF) loaded by a random force, what we nowadays call Miles' equation. Miles' equation is very frequently applied to calculate the r.m.s. response when the SDOF is excited by either random forces or random enforced accelerations or both. Generally, Miles' equation is applied to calculate an approximation of r.m.s. responses of elastic structures (spacecraft, instruments, boxes, etc.) exposed to random loads. White noise random generalized loads are assumed when Miles' equation is derived [2]. The well-known Miles' equation to calculate the r.m.s. of the acceleration $\ddot{x}_{rms}$ (g) of the mass m of the SDOF system is illustrated in Fig. 1.1 and is given by the expression

$$\ddot{x}_{rms} = \sqrt{\frac{\pi}{2} Q f_n W_{\ddot{U}}}, \qquad (1.1)$$

where the amplification (quality) factor $Q = 1/2\zeta$, ζ is the damping ratio, $f_n = \omega_n/2\pi$ (Hz) the natural frequency of the SDOF system, and $W_{\ddot{U}}$ (g^2/Hz) the power (acceleration) spectral density (PSD, ASD) of the random white noise enforced acceleration.

Suppose an instrument is mounted on a spacecraft. When we know the natural frequency f_n of the fundamental vibration mode and the associated PSD of the

J. Wijker, *Miles' Equation in Random Vibrations*, Solid Mechanics and Its Applications 248, https://doi.org/10.1007/978-3-319-73114-8_1

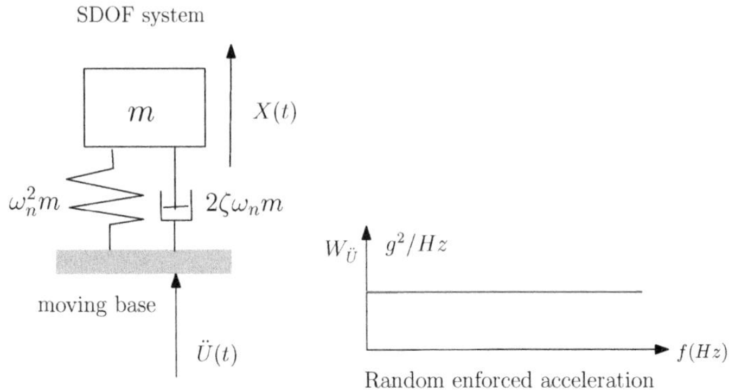

Fig. 1.1 Single degree of freedom system (SDOF), random enforced acceleration

random enforced acceleration $W_{\ddot{U}}(f_n)$ and with Miles' equation, an approximation of the r.m.s. acceleration in that instrument can be made very quickly, before we perform detailed modal and response analyses.

The following assumptions and restrictions are considered in this book. The structure has no random properties and no time varying stiffness and mass, and light damping is assumed. The random process is stationary (does nor change with time) and ergodic (one sample described the random process).

The book about the aspects of Miles' equation contains the following chapters:

In Chap. 2, the r.m.s. responses of a SDOF system loaded by random (white noise) loads and or random (white noise) enforced acceleration, the so-called Miles' equation, are derived applying the Fokker–Planck–Kolgolmorov equation and the 'stochastic dual of the direct method of Lyapunov's method'. The derivation of Miles' equation with the spectral method can be found in standard textbooks about random vibration.

In Chap. 3, the random dynamic response characteristics of structures are approximated by equivalent static analyses. The equivalent static approximation of a random loaded dynamic structure using Miles' equation is only satisfactory when the static deflection under static loads and the vibration modes with associated significant modal effective mass or modal (static) participation factors have similar deformation shapes. In this chapter, two equivalent static methods are discussed: the equivalent static approximation with an acceleration field and the equivalent static approximation with a force field.

Random load factors and mass participation factors are defined and discussed in Chap. 4. 3σ values of the random load factors are applied to analyze in a static manner the strength characteristics (collapse, yield, buckling) of structural elements in spacecraft, instruments, etc. The probability that normally distributed random loads will be beyond the 3σ is Prob $(|X| \geq 3\sigma) \leq 0.0027$. The second aspect discussed in this chapter is the mass participation approach, which represents the random reaction forces at the base when the structure is excited by an enforced random acceleration. With the aid of the mass participation factors, random load factors are derived.

In Chap. 5, the mass participation approach for notching analysis is discussed, in which the modal effective mass and apparent mass in conjunction with Miles' equation are the basic ingredients to determine notched random acceleration input comparing 3σ reaction loads with the reaction loads caused by the QSL design loads.

In Chap. 6, the consideration of the choice performing either a random vibration test or an acoustic test is discussed. Based on the response characteristics of a SDOF system exposed to enforced random vibration or a pressure field, a key factor between the area (surface) exposed to the pressure field and the total mass of the instrument or unit has been derived which supports the decision between a random vibration and acoustic test.

In Chap. 7, a very simple method to predict the r.m.s acceleration of a plate-like structure exposed to random acoustic loads is discussed. The prediction is based on Miles' equation. The analysis approach is a step-by-step procedure to calculate the equivalent static load environment of radiant panels. A more advanced analysis approach is provided in Chap. 8.

In Chap. 8, the approximation of responses of thin-walled shell structures exposed to a random acoustic pressure field is discussed. The dynamic behavior of the shell structure is represented by a SDOF element. The acoustic pressure field is separated in spatial and time domain. Three variations of the spatial distribution of the acoustic pressure field are considered.

The equivalence or similarity of levels and time duration of sinusoidal and random vibration testing is discussed in Chap. 9.

In Chap. 10, the characterization and synthesis of random acceleration vibration specifications are discussed in very detail. The random acceleration vibration test specifications are, in general, enforced accelerations at the interface between spacecraft and subsystems. The random vibrations are mainly induced by the acoustic loads exposed to the spacecraft during launch and performing acoustic tests, representing the launch environment. The measured random accelerations and or similar predictions are broad-banded and shows many peaks. These random acceleration measurements and predictions are converted into more or less equivalent smooth random acceleration vibration test specification, which represent as good as possible the underlying measured and calculated random acceleration responses. The equivalent random acceleration vibration test specification shall not lead to under-testing or significant over-testing of the test item. Several methods are available to reconstruct and characterize in a very structured manner the equivalent random acceleration vibration test specification from the measured and predicted random response data.

In the last Chap. 11, a number of typical examples are numerically worked out and explained.

Appendices have been added to give additional support to methods and techniques when applications of Miles' equation are discussed.

In the Appendix A, the random response of a single degree of freedom (SDOF) system, excited by a random enforced acceleration, is explained. The spectral approach is used. This appendix is added because in a number of chapters this method is applied to compute the power spectral density (PSD) and the r.m.s. value of the response of the mass of the SDOF system.

During flight, the spacecraft is subjected to static and dynamic loads, e.g., quasi-static load, acoustic loads, and random vibrations. Specifications about the quasi-static design loads (QSL) and the manner how the acoustics loads and the random vibrations are prescribed need some more explanation, which will be given in the Appendix B.

In Appendix C, the fast and efficient computation the Gaussian distributed random time series from a PSD is discussed.

In Appendix D, a very efficient method of the computation of the shock response spectrum (SRS) is described. The calculations are done with the discrete approximation of the continuous transfer function.

Rayleigh quotient is frequently applied in many examples and to be applied in problems and therefor discussed in more detail in Appendix E.

In Appendix F, a number of fatigue life prediction methods are discussed, in particular the narrowband method, the Dirlik method, and Steinberg's three-band method.

In all chapter and appendices, illustrative examples are worked out and problems (exercises) are provided.

The last appendix G of this book is an obituary notice of John Wilder Miles.

Problems

1.1 Search for articles on the Internet in which Miles' equation is mentioned.

1.2 Search for articles about John Wilder Miles on the Internet.

References

1. Miles JW (1954) On structural fatigue under random loading. J Aeronaut Sci 21(11):753–762
2. Crandall SH (ed) (1963) Random vibration, vol 2. The M.I.T Press, Cambridge

Chapter 2
Miles' Equation

Abstract The root mean square (r.m.s.) response of a single degree of freedom (SDOF) system when exposed to white noise random excitation can be approximated applying Miles' equation. In this chapter, Miles' equation is derived for a) a random enforced acceleration and b) a random applied force. A further approximation was introduced assuming a more or less flat PSD of acceleration and force in the vicinity of the natural frequency of the SDOF system. The accuracy of the approximation with Miles' equation is investigated, and examples are worked out. Problems with solutions are provided to get more understanding of the limitations of Miles' equation.

Keywords Miles' equation · White noise · Random response analysis · SDOF system

2.1 Introduction

The mean square response of a single degree of freedom (SDOF) system is, mostly, calculated by the classical complex, variable residue theory techniques as discussed in many textbooks about random mechanical vibrations [1–4]. In this section, the mean square responses of a single degree of freedom (SDOF) system excited at the base by enforced acceleration and applied forces are discussed, however, using alternative techniques:

1. The Fokker–Planck–Kolmogorov equation (FPKE) is applied to calculate the mean square responses of the displacement, velocity, and acceleration when SDOF system is exposed by enforced random white noise acceleration [5]
2. The 'Stochastic dual of the direct method of Lyapunov' is applied to solve the mean square responses of the displacement and velocity when the SDOF system is exposed to an applied random white noise force [6, 7]. The mean square solution of the acceleration does not exist because a white noise force is assumed.

White noise processes are assumed for the enforced acceleration and applied force. The mean square (m.s.) responses of a SDOF system are frequently applied to approximate the response characteristics of more complex mechanical systems (continuous and multidegrees of freedom (MDOF) systems). These approximations

J. Wijker, *Miles' Equation in Random Vibrations*, Solid Mechanics and Its Applications 248, https://doi.org/10.1007/978-3-319-73114-8_2

Fig. 2.1 SDOF system, enforced acceleration

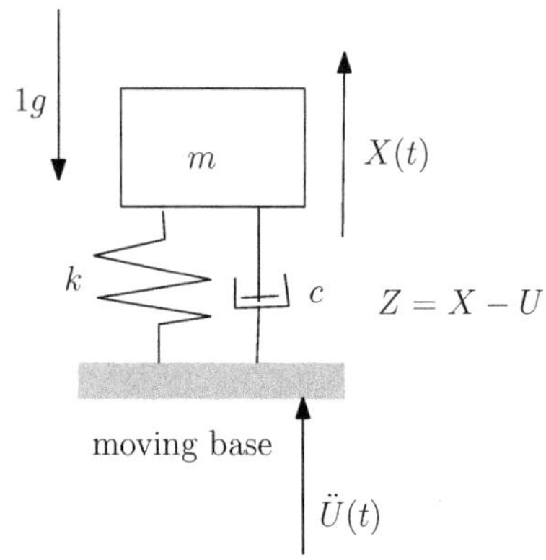

will be illustrated by simple examples. Related to Miles' equation, the 3σ 'static' design approach and the mass participation approach will be discussed.

2.2 SDOF System, Enforced Acceleration

Using Miles' equation [8], the root mean square (r.m.s.) acceleration response of SDOF can be calculated when excited at the base by a random enforced acceleration. This equation is well known to estimate the r.m.s. inertia loads inside a spacecraft structure assuming the one-vibration mode behavior. This Miles' equation will be obtained using the stationary joint probability distribution function (JPDF) of the displacement and velocity of the responses of the SDOF system. The stationary JPDF is the solution of the stationary Fokker–Planck–Kolmogorov equation (FPKE) associated with a Markov process and white noise random enforced acceleration with zero mean and delta correlation [9–12]. The equation of motion of the SDOF dynamic system, illustrated in Fig. 2.1, relative to the base, is given by

$$m\ddot{Z} + c\dot{Z} + kZ = -m\ddot{U}(t), \tag{2.1}$$

where m is the discrete mass, k is the stiffness of the spring element, c is the damping constant, $\ddot{U}(t)$ is the random enforced acceleration. Z is the relative displacement, and X is the absolute displacement.

The absolute acceleration can be obtained as follows:

$$\ddot{X} = -2\zeta\omega_n\dot{Z} - \omega_n^2 Z, \tag{2.2}$$

where the damping ratio is $\zeta = c/c_{cr} = c/2\sqrt{k/m}$, the amplification factor $Q = 1/2\zeta$, and the natural frequency is $\omega_n = \sqrt{k/m}$. The random enforced acceleration

$\ddot{U}(t)$ is a double-sided white noise process with a constant PSD $W_{\ddot{U}}/2$, $-\infty < \omega < \infty$ ((m/s^2)^2/Hz). To solve Eq. (2.1), the FPKE is used.

The state-space equations with the state variables $X_1 = Z$, $X_2 = \dot{Z}$ can be written as

$$\dot{X}_1 = X_2, \tag{2.3}$$

$$\dot{X}_2 = -\omega_n^2 X_1 - 2\zeta\omega_n X_2 - \ddot{U}. \tag{2.4}$$

JPDF $p = p(X_1, X_2)$ is the solution of the stationary FPKE [9]

$$-\frac{\partial}{\partial X_1}(X_2 p) + \frac{\partial}{\partial X_2}[(\omega_n^2 X_1 + 2\zeta\omega_n X_2)p] + \frac{\partial^2}{\partial X_2^2}\left(\frac{W_{\ddot{U}}}{2}p\right) = 0, \tag{2.5}$$

which is

$$p(X_1, X_2) = p(Z, \dot{Z}) = A\exp\left[-\frac{4\zeta\omega_n}{W_{\ddot{U}}}(\dot{Z}^2 + \omega_n^2 Z^2)\right], \tag{2.6}$$

where A is the normalization constant, because $\int_{-\infty}^{\infty} p dz d\dot{z} = 1$. We see that the JPDF p can be written as

$$p(Z_1, Z_2) = p(Z_1)p(Z_2). \tag{2.7}$$

That means that the displacement and the velocity are independent of each other. The probability density functions (PDFs) become

$$p(Z) = 2\sqrt{\frac{\zeta\omega_n^3}{\pi W_{\ddot{U}}}}\exp\left[-\frac{4\zeta\omega_n^3 Z^2}{W_{\ddot{U}}}\right], \tag{2.8}$$

$$p(\dot{Z}) = 2\sqrt{\frac{\zeta\omega_n}{\pi W_{\ddot{U}}}}\exp\left[-\frac{4\zeta\omega_n \dot{Z}^2}{W_{\ddot{U}}}\right]. \tag{2.9}$$

Both PDFs are Gaussian. The mean square values of the displacement and velocity are

$$E\{Z^2\} = \int_{-\infty}^{\infty} Z^2 p(Z) dZ = \frac{W_{\ddot{U}}}{8\zeta\omega_n^3} \tag{2.10}$$

$$E\{\dot{Z}^2\} = \int_{-\infty}^{\infty} \dot{Z}^2 p(\dot{Z}) d\dot{Z} = \frac{W_{\ddot{U}}}{8\zeta\omega_n} \tag{2.11}$$

The mean square values of the absolute acceleration $\ddot{X}$ can now be obtained from (2.2)

$$\begin{aligned} E\{\ddot{X}^2\} &= (2\zeta\omega_n)^2 E\{\dot{Z}^2\} + 2\zeta\omega_n^3 E\{Z\dot{Z}\} + \omega_n^4 E\{Z^2\}, \\ &= \frac{\omega_n W_{\ddot{U}}}{8\zeta}(1 + 4\zeta^2) = \frac{\pi f_n Q W_{\ddot{U}}}{2}(1 + 4\zeta^2). \end{aligned} \tag{2.12}$$

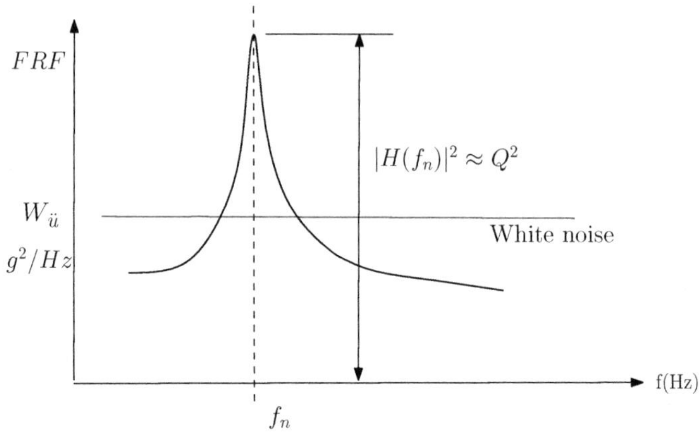

Fig. 2.2 SDOF system response plot (courtesy GSFC FEMCI)

Table 2.1 Random vibration specification

Frequency range (Hz)	PSD(ASD) Levels (g^2/Hz)	Remark
20–100	3 dB/oct.	
100–700	0.4	23.52 G_{rms}
700–2000	−3 dB/oct.	

2.2.1 Further Approximation of Miles' Equation

In spacecraft structures, the modal damping is, in general, low (1–5%). The mean square of the absolute acceleration $\ddot{X}$ (2.12) may be further approximated

$$E\{\ddot{X}^2\} = \frac{\pi f_n Q W_{\ddot{U}}}{2}(1+4\zeta^2) \approx \frac{\pi f_n Q W_{\ddot{U}}}{2} \approx \frac{\pi f_n Q W_{\ddot{U}}(f_n)}{2}, \tag{2.13}$$

where f_n is the natural frequency of the SDOF system and $W_{\ddot{U}}(f_n)$ the spectrum of the random enforced acceleration at f_n. The last approximation is allowed because the frequency response transfer function (FRF) at the resonant frequency is very peaked. This is illustrated in Fig. 2.2. The approximation may be inaccurate when the spectrum of the enforced acceleration varies drastically at the natural frequency f_n.

Equation (2.13) is mostly called Miles' equation. The accuracy of Eq. (2.13) will be illustrated by numerical examples.

The classical random enforced acceleration specification at the base of the SDOF element is provided in Table 2.1. Given this specification, the r.m.s. response of the absolute acceleration $\ddot{X}$ will be computed by the spectral method (Appendix A) and using Miles' equation Eq. (2.13). The value of the modal damping ratio is $\zeta = 0.05$. The results of the computations are shown in Fig. 2.3.

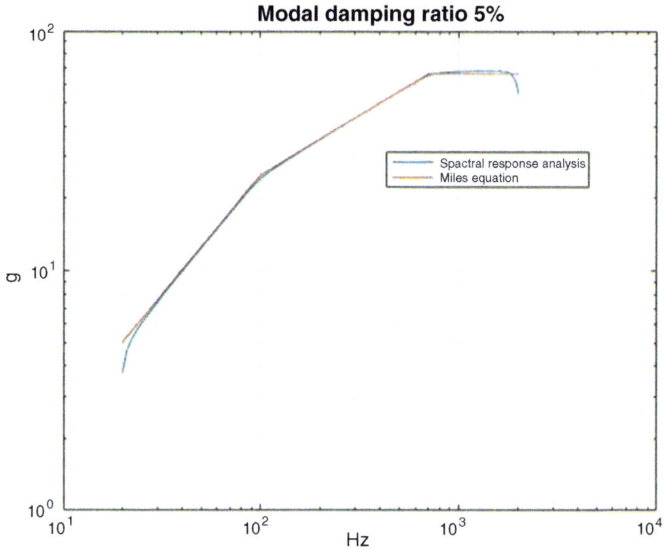

Fig. 2.3 Comparison r.m.s. responses SDOF system: Spectral method and Miles' equation

Table 2.2 Acceleration power spectrum $\ddot{U}$

Frequency range (Hz)	PSD(ASD) Levels (g^2/Hz)	Remark
20–100	3 dB/oct.	
100–200	0.04	
200–300	−5 dB/oct.	4.501 G_{rms}
300–500	0.02	
500–2000	−6 dB/oct.	

Investigation of the plots in Fig. 2.3 shows that the results of the direct response analysis and the results from Miles' equation are very similar, except at the boundaries of the frequency interval, in particular at 20 and 2000 Hz. Below 20 Hz and beyond 2000 Hz, no random acceleration spectrum is specified. That is taken into account at the direct response analysis, but not when Miles' equation is applied. In general, Miles' equation Eq. (2.13) is a very good approximation of Eq. (2.12) when the random enforced acceleration is non-white.

The application of Miles' equation Eq. (2.13) will be investigated when the spectrum of the enforced random acceleration $\ddot{U}$ is applied to the base of the SDOF system (Fig. 2.1). The acceleration spectrum is provided in Table 2.2 and Fig. 2.4.

The square root of the expected values $E(\ddot{X}^2)$ or the r.m.s. values of $\ddot{X}$ for different values of the damping ration ζ are computed applying the spectral method and Miles' equation. The results of the computations are presented in Table 2.3. The difference (error) between both methods is given between the brackets.

The higher the natural frequency f_n of the SDOF system, the higher the error of the results obtained from Miles' equation compared to the spectral method. PSD

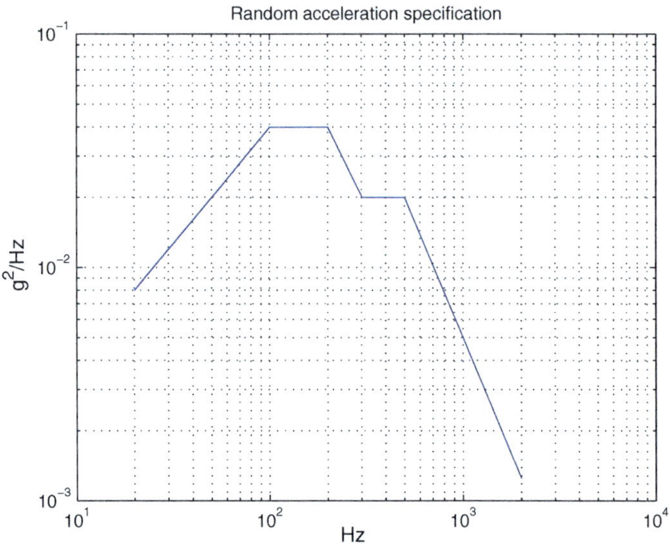

Fig. 2.4 Random acceleration specification

Table 2.3 Comparison spectral response analysis and Miles' equation analysis (error %)

f_n	(Hz)	50	150	250	400	1000
$W_{\ddot{U}}(f_n)$	(g^2/Hz)	0.02	0.04	0.276	0.02	0.005
$\zeta = 0.01$						
$\sqrt{E(\ddot{X})_{spectral}}$	(g)	8.828	21.643	23.360	25.142	20.298
$\sqrt{E(\ddot{X})_{miles}}$	(g)	8.871 (0.5)	21.7108 (0.3)	23.264 (−0.4)	25.066 (−0.3)	19.815 (−2.4)
$\zeta = 0.05$						
$\sqrt{E(\ddot{X})_{spectral}}$	(g)	3.905	9.599	10.618	11.419	9.891
$\sqrt{E(\ddot{X})_{miles}}$	(g)	3.967 (1.6)	9.708 (1.1)	10.404 (−2.0)	11.210 (−1.8)	8.861 (−10.4)
$\zeta = 0.1$						
$\sqrt{E(\ddot{X})_{spectral}}$	(g)	2.755	6.773	7.678	8.284	7.653
$\sqrt{E(\ddot{X})_{miles}}$	(g)	2.805 (1.8)	6.865 (1.4)	7.357 (−4.2)	7.927 (−4.3)	6.266 (−18.1)

below the natural frequency f_n is less taken into account using Miles' equation. Even the results of Miles' equation at locations where PSD is not flat are rather reliable.

Example

A printed circuit board (PCB) in a electronics box to support the electronic function of a scientific instrument with an ozone monitoring space mission has a natural

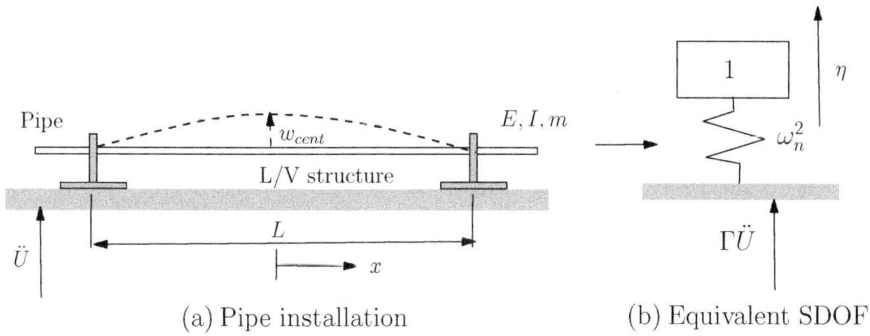

Fig. 2.5 Hydraulic pipe installation

frequency $f_n = 125$ Hz and can be idealized as a SDOF system.The damping ratio is assumed $\zeta = 0.05$. PCB is excited by a random acceleration spectrum specified in Table 2.2. What is the r.m.s. acceleration response of the PCB? The acceleration spectrum at $f_n = 125$ Hz is $W_{\ddot{U}}(f_n) = 0.04\,\text{g}^2/\text{Hz}$. The r.m.s. acceleration (1σ value) can be calculated with Miles' equation

$$\ddot{x}_{rms} = \sqrt{\frac{\pi}{2} f_n Q W_{\ddot{U}}(f_n)} = \sqrt{\frac{\pi}{2} \times 125 \times 10 \times 0.04} = 8.86\ \text{g}.$$

The 3σ value of the r.m.s. acceleration is mostly considered as the peak value $\ddot{x}_{peak}$ or the equivalent static design (peak) load. The peak load is $\ddot{x}_{peak} = 3\text{r.m.s.} = 26.59$ g. The probability that the Gaussian- distributed random acceleration is beyond the 3σ value is Prob$(|\ddot{X}| \geq 3\sigma) \geq 0.0027$.

Example

In Fig. 2.5a, a hydraulic pipe is supported by two supports on a launch vehicle (L/V). The pipe is considered as a simply support uniform beam excited by an enforced acceleration $\ddot{U}$ at the supports. In Fig. 2.5b, the pipe installation is represented by an equivalent SDOF system. w_{cent} represents the maximum displacement, relative to the attachments.

We start with the derivation of the equivalent mass m and spring stiffness k, applying Rayleigh's quotient R. The modal damping ratio is $\zeta = 0.05$ ($Q = 10$). The assumed mode (displacement shape) ϕ is given by [13]

$$\phi(x) = \left(\frac{L^2}{4} - x^2\right)\left(\frac{5L^2}{4} - x^2\right), \quad -\frac{L}{2} \leq x \leq \frac{L}{2}. \tag{2.14}$$

The physical displacement is the product of the generalised coordinate $\eta(t)$ and the assumed mode $\phi(x)$

$$w(x, t) = \eta(t)\phi(x). \tag{2.15}$$

The bending stiffness of the beam is EI and the mass per unit of length m (kg/m), where E (Pa) is Young's modulus and I (m^4) the second moment of area.The Rayleigh quotient [14] is given by the following expression

$$R = \omega_n^2 = \frac{k}{m} = \frac{EI\int_{-\frac{1}{2}L}^{\frac{1}{2}L}\left(\frac{d^2\phi}{dx^2}\right)^2 dx}{m\int_{-\frac{1}{2}L}^{\frac{1}{2}L}\phi^2(x)dx} = \frac{\frac{24EIL^5}{5}}{\frac{31mL^9}{630}} = \frac{97.5484EI}{mL^4}. \tag{2.16}$$

The natural frequency f_n (Hz) is

$$f_n = \frac{\omega_n}{2\pi} = 1.5719\sqrt{\frac{EI}{mL^4}}. \tag{2.17}$$

When PSD of the enforced acceleration $W_{\ddot{U}}(f_n)$ (g^2/Hz) is known, the r.m.s. acceleration response of the SDOF system can be calculated with Miles' equation.

$$\ddot{\eta}_{rms} = |\Gamma|\sqrt{\frac{\pi}{2}f_n Q W_{\ddot{U}}(f_n)}\ \mathrm{g}, \tag{2.18}$$

where Γ is modal participation factor given by

$$\Gamma = \frac{-m\int_{\frac{-1}{2}L}^{\frac{1}{2}L}\phi(x)dx}{m\int_{\frac{-1}{2}L}^{\frac{1}{2}L}\phi^2(x)dx} = -\frac{126}{31L^4}. \tag{2.19}$$

The r.m.s. displacement and acceleration response of the pipe are as follows:

$$\begin{aligned} w_{rms}(x) &= \frac{\ddot{\eta}_{rms}}{\omega_n^2}\phi(x), \\ \ddot{w}_{rms}(x) &= \ddot{\eta}_{rms}\phi(x). \end{aligned} \tag{2.20}$$

The r.m.s. values of the bending moment $M_{b,rms}(x)$ and shear force $D_{s,rms}(x)$ are given by

$$\begin{aligned} M_{b,rms}(x) &= \frac{\ddot{\eta}_{rms}}{\omega_n^2}EI\left|\frac{d\phi^2(x)}{dx^2}\right|, \\ D_{bs,rms}(x) &= \frac{\ddot{\eta}_{rms}}{\omega_n^2}EI\left|\frac{d\phi^3(x)}{dx^3}\right|. \end{aligned} \tag{2.21}$$

The bending moment at the mid-span of the pipe $M_{b,rms}(0)$ and the shear force at the supports $D_{bs,rms}(\pm L/2)$ are given by he following expressions:

$$M_{b,rms}(0) = \frac{\ddot{\eta}_{rms}}{\omega_n^2} 3EIL^2,$$
$$D_{bs,rms}\left(\pm\frac{L}{2}\right) = \frac{\ddot{\eta}_{rms}}{\omega_n^2} 12EIL. \quad (2.22)$$

PSD of the enforced acceleration is $W_{\ddot{U}}(f) = 2\ \mathrm{g}^2/\mathrm{Hz}$ between 20 and 200 Hz. The steel pipe has a length $L = 450$ mm between the supports. The outer diameter $D = 8$ mm and the inner diameter $d = 6$ mm. Young's modulus of steel is $E = 210$ GPa and the density $\rho_s = 7800\ \mathrm{kg/m^3}$. The density of the oil is $\rho_o = 900\ \mathrm{kg/m^3}$. The second moment of area of a pipe is $I = \pi/64(D^4 - d^4)$ (m^4), and the first moment of area with respect to the center is $Q_s = \pi/48(D^3 - d^3)$ (m^3). The moment of resistance $W_b = 2I/D$ (m^3). The mass per unit of length of the pipe is $m = \rho_s\pi/4(D^2 - d^2) + \rho_o\pi/4d^2$.

The calculated natural frequency of the simply supported pipe is $f_n = 134.9040$ Hz.

The bending stress at the mid-span and the shear stress at supports are

$$\sigma_{b,rms} = \frac{M_{b,rms}(0)}{W_b} = 3.6418 \times 10^7\ \mathrm{Pa},$$
$$\tau_{s,rms} = \frac{D_{bs,rms}\left(\pm\frac{L}{2}\right) Q_s}{I(D-d)} = 7.0553 \times 10^5\ \mathrm{Pa}. \quad (2.23)$$

The mean values of the stresses are zero. Therefor the r.m.s. values are the 1σ values of the stresses. 3σ values of the stresses are used for strength calculations.

2.3 Force-Loaded SDOF System

The 'Stochastic dual of the direct method of Lyapunov' [6, 15] will be applied to calculate the response characteristics of a single degree of freedom system loaded by a stationary white noise random force. The SDOF system is shown in Fig. 2.6.

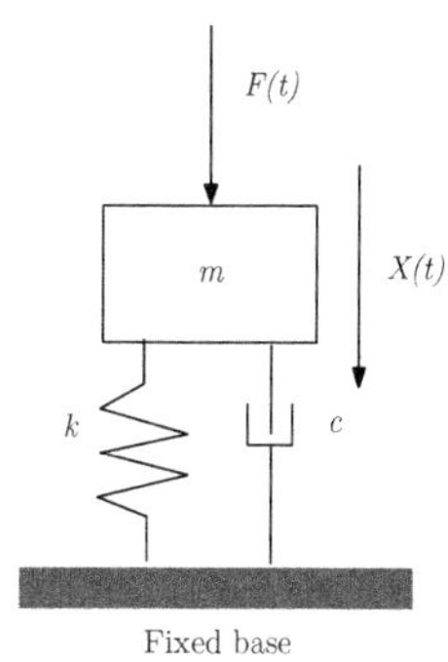

Fig. 2.6 SDOF system loaded by a force

The discrete mass m is supported by a damper with a damping constant c, and a spring with spring stiffness k. The absolute displacement is indicated by $X(t)$. The equation of motion of the single degree of freedom (SDOF) system is given by

$$m\ddot{X}(t) + c\dot{X}(t) + kX(t) = F(t), \tag{2.24}$$

Equation (2.24) can be written

$$\ddot{X}(t) + 2\zeta\omega_n\dot{X}(t) + k\omega_n^2(t) = \frac{F(t)}{m}. \tag{2.25}$$

Introducing the state-variables $Y_1 = X$ and $Y_2 = \dot{X}$, the equation of motion (2.25) can be expressed in state equations

$$\dot{Y}_1 = Y_2 \tag{2.26}$$

$$\dot{Y}_1 == 2\zeta\omega_n Y_2 - \omega_n^2 Y_1 + \frac{F(t)}{m}, \tag{2.27}$$

or

$$\{\dot{Y}\} = [A]\{Y\} + \{B\}F(t), \tag{2.28}$$

where the matrix $[A]$ is the state (Jacobian [16]) matrix and the vector $\{B\}$ is the filter vector, both given by

$$[A] = \begin{bmatrix} 0 & 1 \\ -2\zeta\omega_n & -\omega_n^2 \end{bmatrix}, \ \{B\} = \begin{pmatrix} 0 \\ \frac{1}{m} \end{pmatrix}. \tag{2.29}$$

The applied force $F(t)$ is a double-sided white noise random force with zero mean and correlation function $E\{F(t)F(t+\tau)\} = \delta(\tau)W_F/2$. The correlation matrix $[P]$ of the responses, displacement, and velocity is given by

$$[P] = \begin{bmatrix} E\{Y_1^2\} & E\{Y_1Y_2\} \\ E\{Y_2Y_1\} & E\{Y_2^2\} \end{bmatrix}, \tag{2.30}$$

where $E\{Y_1Y_2\} = E\{Y_2Y_1\} = 0$ for a stationary process. The mean square response of the displacement and velocity can be obtained by the following Lyapunov (Riccatti) equation

$$[A][P] + [P][A]^T = -\{B\}\frac{W_F}{2}\{B\}^T. \tag{2.31}$$

Equation (2.31) can be rearranged as [15]

$$\begin{bmatrix} -\omega_n^2 & 1 \\ 0 & -2\zeta\omega_n \end{bmatrix} \begin{pmatrix} E\{Y_1^2\} \\ E\{Y_2^2\} \end{pmatrix} = \frac{-W_F}{2m^2}. \tag{2.32}$$

Finally, we can write the following solutions for the mean square responses

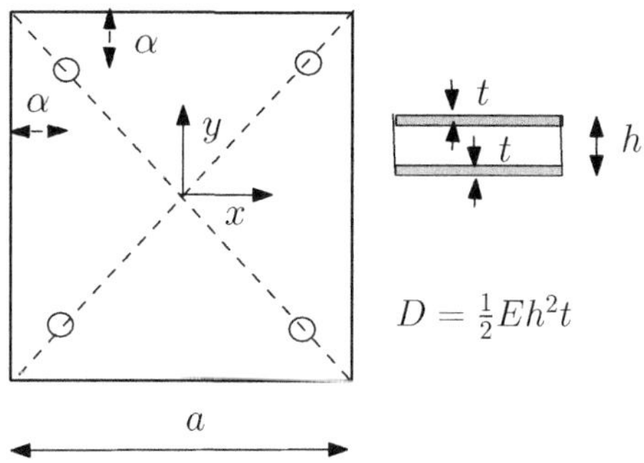

Fig. 2.7 Square solar panel supported symmetrically on four points

$$E\{X^2\} = \frac{W_F}{8\zeta\omega_n^3 m^2} = \frac{W_F}{8\zeta(2\pi f_n)^3 m^2}, \tag{2.33}$$

$$E\{\dot{X}^2\} = \frac{W_F}{8\zeta\omega_n m^2} = \frac{W_F}{8\zeta(2\pi f_n) m^2}, \tag{2.34}$$

where the cyclic frequency is $f_n = \omega_n/2\pi$. The mean square of the acceleration $E\{\ddot{X}^2\}$ does not exist because $E\{F^2\}$ does not exist for a white noise process.

When the applied load a white noise random pressure field with a zero mean and a correlation function $E\{p(t)p(t+\tau)\} = \delta(\tau)W_p/2$, then (2.33) and (2.34) become

$$E\{X^2\} = \frac{A^2 W_p}{8\zeta\omega_n^3 m^2} = \frac{A^2 W_p}{8\zeta(2\pi f_n)^3 m^2}, \tag{2.35}$$

$$E\{\dot{X}^2\} = \frac{A^2 W_p}{8\zeta\omega_n m^2} = \frac{A^2 W_p}{8\zeta(2\pi f_n) m^2}, \tag{2.36}$$

where A is the area.

The mean square value of the spring force in the spring of the SDOF system becomes, when $Q = 1/2\zeta$

$$k^2 E\{X^2\} = \frac{(2\pi f_n)^4 m^2 A^2 W_p}{8\zeta(2\pi f_n)^3 m^2} = A^2 \frac{\pi}{2} f_n Q W_p, \tag{2.37}$$

which looks similar to the Miles' equation of the approximation of mean square of the acceleration $E\{\ddot{X}^2\}$ in case of enforced acceleration Eq. (2.12).

Example

A square solar sandwich panel is supported at the four discrete points and is illustrated in Fig. 2.7.

The sandwich panel has Al-alloy face sheets with thickness $t = 0.3$ mm, Young's modulus of the face sheets is $E = 7 \times 10^{10}$ Pa, $\nu = 0.33$, and the mass per unit of area of the panel is $m = 2\,\text{kg/m}^2$. The height of the core is $h = 20$ mm. The shear/bending stiffness of the core shall be neglected. The length/width is $a = 1.5$ m and $\alpha = 0.2$ m. The panel is dynamically loaded by an uniform random pressure field P with $SPL = 135\,\text{dB}$ ($p_{ref} = 2 \times 10^{-5}\,\text{Pa}$) at the octave band frequency $f_c = 63\,\text{Hz}$. Calculate the r.m.s. displacement at center of the panel ($x = 0$, $y = 0$). The deflection of the panel $W(x, y, t)$ can be written as

$$W(x, y, t) = \eta(t)\Phi(x, y), \tag{2.38}$$

where $\eta(t)$ is the generalized coordinate and $\Phi(x, y)$ is the assumed mode taken from [17]

$$\Phi(x, y) = \left(2 - \frac{2^2k^2x^2}{a^2} - \frac{2^2k^2y^2}{a^2}\right), \tag{2.39}$$

with $k = a/(a - 2\alpha)$.

At first the natural frequency ω_n^2 will be estimated with the usage of Rayleigh's quotient. The strain energy U is given by

$$U = \frac{D}{2}\int_{-\frac{a}{2}}^{\frac{a}{2}}\int_{-\frac{a}{2}}^{\frac{a}{2}}\left[\left(\frac{\partial^2\Phi}{\partial x^2}\right)^2 + \left(\frac{\partial^2\Phi}{\partial y^2}\right)^2 + 2\nu\frac{\partial^2\Phi}{\partial x^2\partial y^2} + 2(1-\nu)\left(\frac{\partial^2\Phi}{\partial x\partial y}\right)^2\right]dxdy, \tag{2.40}$$

and the 'kinetic' energy T^* is

$$T^* = \frac{m}{2}\int_{-\frac{a}{2}}^{\frac{a}{2}}\int_{-\frac{a}{2}}^{\frac{a}{2}}\Phi^2dxdy = \frac{m_g}{2}, \tag{2.41}$$

where m_g is the generalised mass. The approximation of the natural frequency ω_n^2 is associated with the assumed mode $\Phi(x, y)$ and can be obtained with

$$\omega_n^2 = \frac{U}{T^*} = \frac{1440k^4D}{\left(7a^4k^4 - 30a^4k^2 + 45\,a^4\right)m}, \quad f_n = 62.4400\text{ Hz}. \tag{2.42}$$

The calculated natural frequency f_n in Eq. (2.42) is confirmed by the results presented in [18], namely $f_n = 62.3018\,\text{Hz}$.

The damped equation of motion associated with the assumed mode $\Phi(x, y)$ can be expressed in $\eta(t)$ as

$$\ddot{\eta}(t) + 2\zeta\omega_n\dot{\eta}(t) + \omega_n^2\eta(t) = \Gamma P(t), \tag{2.43}$$

where $P(t)$ is the random pressure and the modal participation factor Γ can be obtained with the following expression

$$\Gamma = \frac{\int_{-\frac{a}{2}}^{\frac{a}{2}}\int_{-\frac{a}{2}}^{\frac{a}{2}}\Phi(x, y)dxdy}{m_g} = \frac{15k^2 - 45}{14k^4 - 60k^2 + 90} = 0.3187. \tag{2.44}$$

The generalised mass m_g is

$$m_g = 2T^* = 0.0222(28a^2k^4 - 120a^2k^2 + 180a^2)m = 5.3676. \tag{2.45}$$

The r.m.s. of η can be calculated with

$$\eta_{rms} = |\Gamma| \frac{A_p}{m_g} \frac{1}{(2\pi f_n)^2} \sqrt{\frac{\pi}{2} f_n Q W_p} = 4.5805 \times 10^{-4}, \tag{2.46}$$

where A_p the the area of the panel, $Q = 10$ is the amplification factor and PSD of the pressure field (see Appendix B)

$$W_p(f_c) = p_{ref}^2 \frac{10^{\frac{SPL}{10}}}{0.5\sqrt{2}fc} = 2.8394 \times 10^2 \ \text{Pa}^2/\text{Hz}. \tag{2.47}$$

The r.m.s. displacement at the center of the panel is $w_{rms}(0,0) = \eta_{rms}\Phi(0,0) = 9.1611 \times 10^{-4}$ m. The bending moments are constant and can be derived as follows

$$\begin{aligned} |M_x| = |M_y| &= \eta_{rms} D \left(\frac{\partial^2 \phi(x,y)}{\partial x^2} + \nu \frac{\partial^2 \phi(x,y)}{\partial y^2} \right) = \eta_{rms} D \left(\frac{\partial^2 \phi(x,y)}{\partial y^2} + \nu \frac{\partial^2 \phi(x,y)}{\partial x^2} \right) \\ &= \eta_{rms} \frac{D\left(8k^2\nu + 8k^2\right)}{a^2} = 3.6932 \times 10^4 \ \text{Nm}. \end{aligned} \tag{2.48}$$

The bending stress (stress in the face sheets) is

$$\sigma_b = M_x/W = 0.5 M_x h/I = 2.8195 \times 10^6 \ \text{Pa}.$$

2.4 Chapter Summary

The r.m.s. response of SDOF can be approximated by Miles' equation. In this chapter, Miles' equation was derived for a) a random enforced acceleration and b) a random applied force. A further approximation was introduced assuming a more or less flat PSD of acceleration and force in the vicinity of the natural frequency f_n of the SDOF system. The accuracy of the approximation with Miles' equation was investigated, and examples are worked out.

Problems

2.1 Abdelal et al. proposed in their book [19] a frequency- dependent approximation of the damping ratio ζ

$$\zeta = \frac{1}{10 + 0.05 f_n}, \tag{2.49}$$

where f_n (Hz) is the spacecraft (satellite) natural frequency. Rewrite the second part of Eq. (2.13) substituting $Q = 1/2\zeta$ given in Eq. (2.49).

Answer: $Q = 5 + 0.025 f_n$.

2.2 The equation of motion of a SDOF system with Rayleigh damping is excited by a random enforced acceleration $\ddot{U}$. $Z = X - U$ is the relative motion. The equation of motion is given by

$$m\ddot{X} + \alpha m\dot{Z} + \beta k\dot{Z} + kZ = 0. \tag{2.50}$$

Perform or answer the following assignments/questions:

- Rewrite Eq. (2.50) in the relative motion of Z.
- Define the equivalent damping ratio ζ, $(\omega_n^2 = k/m)$.
- Has the Rayleigh damping any consequence on Miles equation Eq. (2.13).

Answers: $\ddot{Z} + (\alpha + \beta\omega_n^2)\dot{Z} + \omega_n^2 Z = -\ddot{U}$, $\zeta = 1/2(\alpha/\omega_n + \beta\omega_n)$.

2.3 Derive Miles' equation with the aid of spectral method [5, 20, 21]. The SDOF system is excited by a random enforced acceleration (see Fig. 2.1).

2.4 The SDOF system in Fig. 2.6 is loaded by a random white noise force with a PSD W_F. Show that the mean square acceleration $E(\ddot{X}^2)$ does not exist (Fig. 2.8).

2.5 The SDOF system with the structural damping force jkg is shown in Fig. 2.10. The equation of motion can only be presented in the frequency domain ω

$$\begin{aligned} \ddot{X}(\omega) + \omega_n^2(1 + jg)Z(\omega) &= 0 \\ \ddot{Z}(\omega) + \omega_n^2(1 + jg)Z(\omega) &= -\ddot{U}(\omega), \end{aligned} \tag{2.51}$$

where the natural frequency is $\omega_n^2 = \sqrt{k/m}$. The relative displacement $Z(\omega)$ can be expressed as follows

$$Z(\omega) = -H(\omega)\ddot{U}(\omega), \tag{2.52}$$

where the frequency transfer function $H(\omega)$ is given by

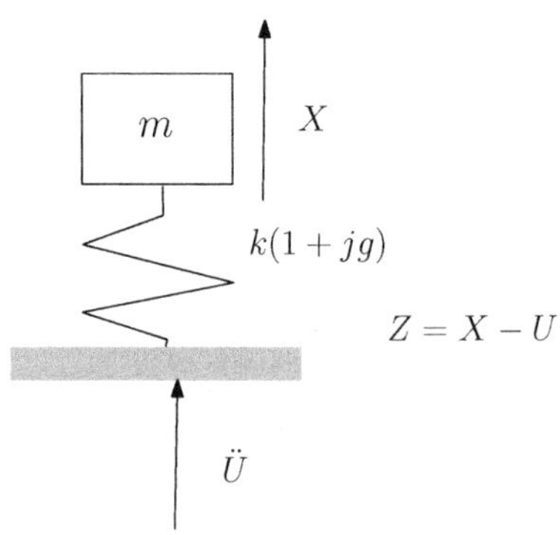

Fig. 2.8 SDOF with structural damping $(j = \sqrt{-1})$

$$H(\omega) = \frac{1}{-\omega^2 + \omega_n^2(1 + jg)} = \frac{1}{\omega_n^2} \frac{1}{[(1 - r^2) + jg]} = \frac{1}{\omega_n^2} H(r), \tag{2.53}$$

where $r = \omega/\omega_n$. To obtain the mean square values of $\ddot{Z}$, $\ddot{z}_{ms}$, the following integral must be solved

$$z_{ms} = \frac{S_{\ddot{U}}}{2\pi\omega_n^3} \int_{-\infty}^{\infty} |H(r)|^2 dr, \tag{2.54}$$

where the two-sided white noise PSD function $S_{\ddot{U}}$ can be replaced by the one-sided white noise PSD function $W_{\ddot{U}} = 2S_{\ddot{U}}$ (g^2/Hz). Equation (2.54) can be replaced by

$$z_{ms} = \frac{W_{\ddot{U}}}{2\pi\omega_n^3} \int_0^{\infty} |H(r)|^2 dr. \tag{2.55}$$

The mean square value of the absolute acceleration $\ddot{x}_{ms}$ is given by

$$\ddot{x}_{ms} = \omega_n^4(1 + g^2) z_{ms}. \tag{2.56}$$

Use the method 'The residue Theorem Evaluation of Integrals' described in [22] to solve Eq. (2.55). Perform the following assignments:

- Prove that $I = \int_0^{\infty} |H(r)|^2 dr = \frac{\pi}{2\sqrt{2}} \frac{\sqrt{(g^2+1)^{\frac{1}{2}}+1}}{(g^2+1)^{\frac{1}{2}} g}$.
- Show numerically the integral $I = \int_0^{\infty} |H(r)|^2 dr \approx \pi/2g$ (see Fig. 2.9). Hint: use, e.g., Matlab/Octave command 'residue'.
- Derive an expression for the absolute mean square acceleration $\ddot{x}_{ms}$.

2.6 The SDOF system with the structural damping force jkg is exposed to a random force F is shown in Fig. 2.10. The one-sided PSD function of the random white noise force is given by W_F. The equation of motion can only be presented in the frequency domain ω:

$$\ddot{X}(\omega) + \omega_n^2(1 + jg)X(\omega) = F(\omega). \tag{2.57}$$

Derive the expression for the mean square value of the displacement X, x_{ms}, using the results from problem 2.5 and compare with Eq. (2.33)
Answer: $x_{ms} = W_F/(4g\omega_n^3 m^2)$.

2.7 A SDOF system with a natural frequency $f_n = 120$ Hz is subjected to a white noise enforced acceleration spectrum $W_{\ddot{U}} = 0.04\,g^2$/Hz. The r.m.s. relative acceleration $\ddot{z}_{rms} = 8.6832$ g. Calculate the amplification factor Q and associated damping ratio ζ.
Answer: $Q = 10$, $\zeta = 0.05$.
Discuss a more general method to estimate the damping ratio of a spacecraft structure.

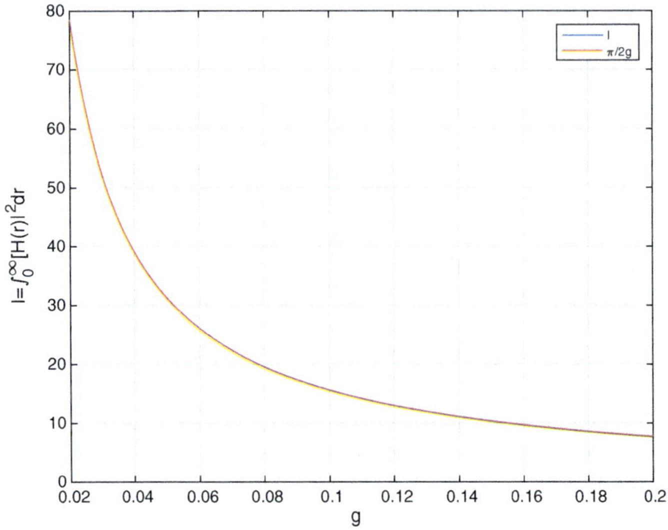

Fig. 2.9 Numerical calculation of $I = \int_0^\infty |H(r)|^2 dr$

Fig. 2.10 SDOF with structural damping ($j = \sqrt{-1}$)

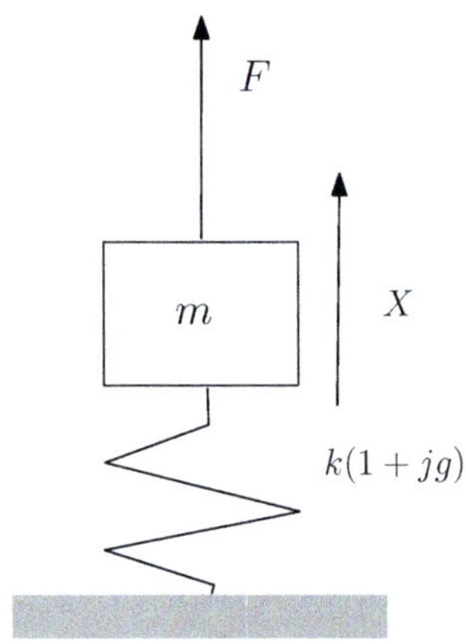

2.8 A simply supported circular plate with radius $r = a$, bending stiffness D, and mass per unit of area $m = \rho t$ supports a discrete mass of $M = 5\,\text{kg}$ in the center of the plate ($r = 0$). The natural frequency is given by

$$\omega_n^2 = \frac{6380010\,\pi\,D}{937445a^2M + 256362\pi\,a^4m}.$$

The supporting plate is made of Al-alloy (E=70 Gpa, $\rho = 2700\,\text{kg/m}^3$, $\nu = 0.33$). The minimum natural frequency is $f_n = 125\,\text{Hz}$. Design the supporting plate with the variables: the thickness t and the radius a. The second moment of area is

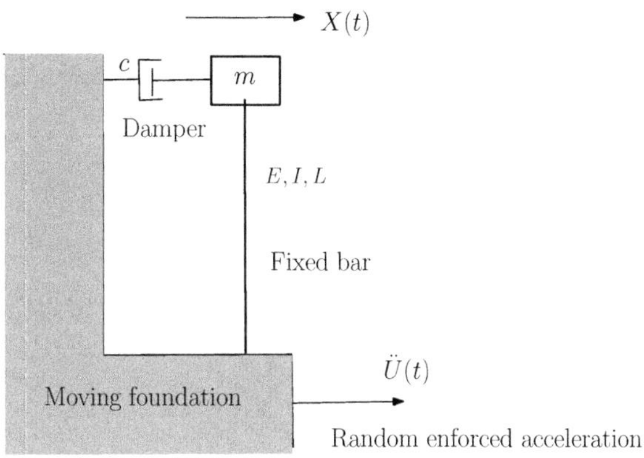

Fig. 2.11 Cantilevered beam excited at the base

$I = t^3/(12(1-\nu^2))$. PSD of the random enforced acceleration $W_{\ddot{U}}(f_n) = 0.1\,\text{g}^2/\text{Hz}$. Calculate approximately the r.m.s. acceleration of the discrete mass using Miles' equation.

2.9 This problem is taken from [1] Chap. 1. A beam is fixed to a moving base with an enforced acceleration $\ddot{U}(t)$. The mass is connected with a damper c. The beam has a length L, a second moment of area I, and Young's modulus of the beam material is E (Fig. 2.11).

The mass, beam and damper system can be transformed into a SDOF system which can be described with the following equation of motion

$$m\ddot{X}(t) + 2\zeta\omega_n m\dot{X}(t) + \omega_n^2 mX(t) = 0. \tag{2.58}$$

Calculate the following parameters:

- the natural frequency ω_n.
- the damping ratio ζ ($c = 2\zeta\omega_n m$).
- when PSD of the enforced white noise acceleration is given by $W_{\ddot{U}}$, calculate the r.m.s. acceleration of the mass $\ddot{x}_{rms}$ using Miles' equation.

Answers: $\omega_n^2 = 3EI/mL^3$, $\zeta = c/2\omega_o m$.

2.10 A simply supported circular sandwich panel in the service module of a S/C supports a number of boxes and cable harness. The radius is $r = a = 1$ m. The panel is shown in Fig. 2.12. The total mass of the boxes plus cable harness $M_{tot} = 100$ kg. One box with a mass of $M_{box} = 30$ kg is mounted to the sandwich panel through four bolts fixed in an insert potting with diameter $\Phi = 40$ mm.

The sandwich panel consists of Al-alloy face sheets with a thickness $t = h/50$ mm. Design a simply supported sandwich panel with a lowest natural frequency $f_n = 75$ Hz. Young's modulus of the Al-alloy $E = 70$ GPa and the density $\rho_a = 2700\,\text{kg/m}^3$. The density of the honeycomb core is $\rho_c = 50\,\text{kg/m}^3$. The total

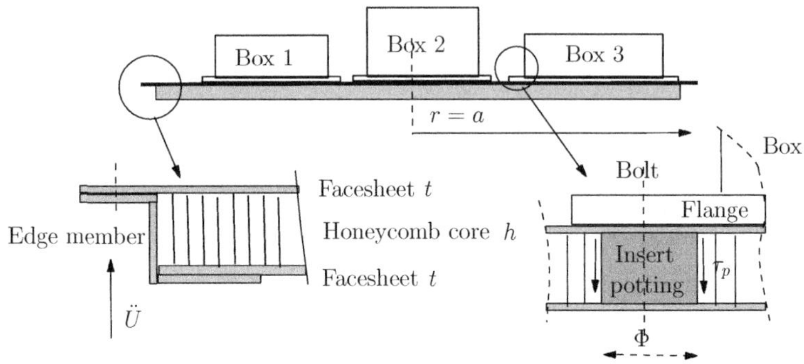

Fig. 2.12 S/C panel supporting equipment

mass of the boxes and cable harness may be uniformly smeared over the circular panel. The second moment of area of the sandwich panel is $I = 0.5h^2t\,(\text{m}^3)$.

The lowest natural frequency (Hz) of the simply supported circular panel can be obtained by the following equation [23]

$$f_n = \frac{4.977}{2\pi}\sqrt{\frac{EI}{ma^4}} = 0.7921\sqrt{\frac{EI}{ma^4}}, \tag{2.59}$$

where m (kg/m^2) is the uniformly mass per unit of area.
Answers: $h = 83.21\,\text{mm}$, $t = 1.66\,\text{mm}$, $m = 44.98\,\text{kg/m}^2$, $f_n = 75.00\,\text{Hz}$.

PSD of the random enforced acceleration at f_n is $W_{\ddot{U}}(f_n) = 0.1\,\text{g}^2/\text{Hz}$ and the damping ratio is $\zeta = 0.05$. Calculate the r.m.s. acceleration G_{rms} of the sandwich panel using Miles' equation.
Answers: $G_{rms} = 10.85\,\text{g}$, $G_{rms} = 106.48\,\text{m/s}^2$.

Calculate the 3σ shear stress τ_p (Pa) around the potting of one of the four bolts of the box $M_{box} = 30\,\text{kg}$.

$$\tau_p = \frac{3M_{box}G_{rms}}{4\pi\Phi h} \tag{2.60}$$

Answers: $\tau_p = 0.229\,\text{MPa}$.

2.11 This problem is based on an application training about random vibration given by Barry Controls, Hutchingson Aerospace & Industry. Determine the required 3σ displacement x for a mounting of an electronic package with the following characteristics:

- Natural frequency $f_n = 50\,\text{Hz}$.
- Amplification factor $Q = 10$.
- Random enforced acceleration $W_{\ddot{U}} = 0.06\,\text{g}^2/\text{Hz}$.

Perform the following assignments:

- Calculate G_{rms}.

- Calculate x_{rms}, $g = 9.81\,\text{m/s}^2$.
- Calculate 3σ value of displacement x.

Answers: $G_{rms} = 6.140\,\text{g}$, $x_{rms} = 6.8232 \times 10^{-04}\,\text{m}$, $3\sigma = 2.05\,\text{mm}$.

References

1. Crandall SH (ed) (1963) Random vibration, vol 2. The M.I.T Press, Cambridge
2. Elishakoff I (1983) Probabilistic methods in the theory of structures. Wiley, New York. ISBN 0 471-87572
3. Lutes LD, Sarkani S (2004) Random vibrations, analysis of structural and mechanical systems. Elsevier, Amsterdam. ISBN 0-7506-7765-1
4. Newland DE (1993) An introduction to random vibrations, spectral and wavelet analysis. Longman scientific technical. ISBN 0-582-21584-6
5. Wijker JJ (2009) Random vibrations in spacecraft structures design, theory and applications, vol 165. Solid mechanics and its applications (SMIA). Springer, Berlin. ISBN 978-90-481-2727-6
6. de la Fuente E (2008) An efficient procedure to obtain exact solutions in random vibration analysis of linear structures. Eng Struct 30:2981–2990
7. Gersch W (1969) Mean-square responses in structural systems. J Acoust Soc Am 48(1): 403–412 (Part 2)
8. Miles JW (1954) On structural fatigue under random loading. J Aeronaut Sci 21(11):753–762
9. Caughey TK (1963) Derivation and application of the Fokker–Planck equation to discrete dynamic systems subjected to white random excitation. J Acoust Soc Am 35(11):1683–1692
10. Sun J (2006) Stochastic dynamics and control. Elsevier, Amsterdam. ISBN 0-444-52230
11. To CWS (2000) Nonlinear random vibration. Swets and Zeitlingerr, Amsterdam. ISBN 90-265-1637-1
12. Wax N (1954) Selected papers on noise and stochastic processes. Dover Publications, New York. ISBN 0-486-60262
13. Temple G, Bickley WG (1956) Rayleigh's principle and applications to engineering. Dover Publications, New York
14. Preumont A (2013) Twelve lectures on structural dynamics, vol 198. Solid mechanics and its applications. Springer, Berlin. ISBN 978-94-007-6382-1
15. Gersch W (1970) Mean-square responses in structural systems. J Acoust Soc Am 48(2):403–413
16. Inan E (ed) (2008) Random vibration of a simple oscillator under different excitation. Springer Science + Bussiness Media B.V, 1–3 Feb 2008
17. Johns DJ, Nagaraj VT (1968) On the fundatmental frequency of square plate symmetrically supported at four points. J Sound Vib 10(3):404–410
18. Raju IS, Amba-Rao CL (1982) Free vibration of a square plate symmetrically supported at for points on the diagonals. J Sound Vib 90(2):291–297
19. Addelal GF, Abuelfoutouh N, Gad AH (2013) Finite element analysis for satellite structures, applications to their design, manufacture and testing. Springer, Berlin. ISBN 978-1-4471-4637-7 (eBook)
20. Maymon G (1998) Some engineering applications in random vibrations and random structures. Progress in astronautics and aeronautics, vol 178. American institute of aeronatics and astronautics. ISBN 1-56347-258-0
21. Wirsching PH, Paez TL, Ortiz H (1995) Random vibrations, theory and practice. Wiley, New York. ISBN 0-471-58579-3
22. Spiegel MR, Lipschutz S, Schiller JJ, Spellman D (2009) Complex variables with an introduction to conforml mapping and its applications, 2nd edn. Schaum's outline series. ISBN 978-0-07-161570-9
23. Blevins RD (1995) Formulas for natural frequency and mode shape. Krieger Publishing Company, New York. ISBN 0-89464-894-2

Chapter 3
Static Equivalent of Miles' Equation

Abstract The replacement of dynamic random response analysis by a simple approximate quasi-static analysis in combination with Miles' equation is discussed in this chapter. Equivalent static acceleration and force fields are considered. A procedure to perform an equivalent finite element (stress) analysis is presented. Examples are given to illustrate this equivalent quasi-static approach. Problems with solutions are provided to gain more insight in using Miles' equation in quasi-static applications.

Keywords Equivalent static loads · Miles' equation · Finite element (stress) analysis

3.1 Introduction

In some cases, more complex dynamic systems may be represented by a simple SDOF system and random response characteristics then can be achieved in a very straightforward manner in combination with Miles' equation. This will be presented for dynamic systems loaded by enforced random acceleration or loaded by random (pressure) loads.

The equivalent static approximation of a random loaded dynamic structure using Miles' equation is only satisfactory when the static deflection under static loads and the vibration modes with associated significant modal effective mass or modal (static) participation factors [1] have similar deformation shapes [2].

In this chapter, two equivalent static methods are discussed:

- The equivalent static acceleration field.
- The equivalent static force field.

In this chapter, a number of illustrative examples are presented and discussed.

J. Wijker, *Miles' Equation in Random Vibrations*, Solid Mechanics and Its Applications 248, https://doi.org/10.1007/978-3-319-73114-8_3

3.2 Equivalent Static Acceleration Field

The approximation of Miles' equation (2.13) is given by

$$\ddot{x}_{rms} \approx \sqrt{\frac{\pi}{2} f_n Q W_{\ddot{U}}(f_n)}.$$

The r.m.s. value of the displacement $x_{rms} = \ddot{x}_{rms}/(2\pi f_n)^2$, thus

$$x_{rms} = \frac{1}{(2\pi f_n)^2}\sqrt{\frac{\pi}{2} f_n Q W_{\ddot{U}}(f_n)}. \tag{3.1}$$

The static displacement x_{stat} of the discrete mass in Fig. 2.1 under a gravitational field of 1 g is given by

$$x_{stat} = \frac{mg}{k} = \frac{g}{(2\pi f_n)^2}. \tag{3.2}$$

Equation 3.2 can be written as

$$x_{rms} = x_{stat}|_{1g}\sqrt{\frac{\pi}{2} f_n Q W_{\ddot{U}}(f_n)}. \tag{3.3}$$

The interpretation of Eq. 3.3 is that the static displacement field or stress distribution of the structure under the influence of the gravitational field of $1g$ multiplied by Miles' equation is an approximation of the r.m.s. displacement and stress distribution of that structure excited by an enforced random acceleration $\ddot{U}$.

To illustrate the static equivalent of enforced acceleration, examples will be worked out and discussed. The quasi-static solutions are compared with random dynamic response analyses:

- A fixed-free (cantilevered) bending beam with tip mass excited at the base by a random enforce acceleration.

3.3 Fixed-Free Beam

The next example is a fixed-free (cantilevered) beam with bending stiffness EI, length L_b, and mass m_b per unit length. The mass at the tip is m. The deflection w is relative to the base. The displacement X is the absolute displacement of the tip mass. The beam mass system is illustrated in Fig. 3.1.

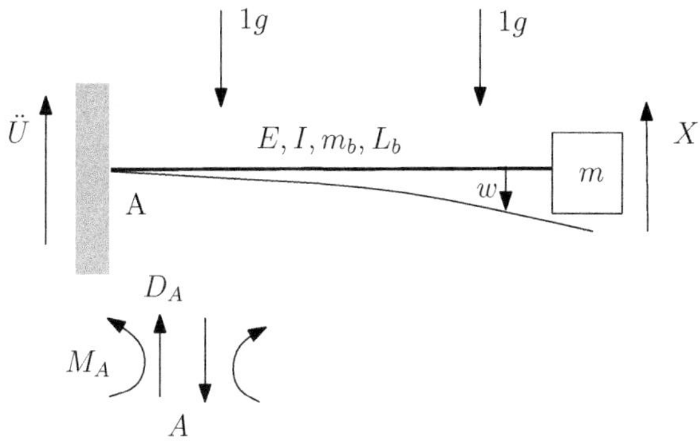

Fig. 3.1 Fixed-free beam with tip mass excited at the base A

Table 3.1 PSD enforced acceleration $W_{\ddot{U}}$

Frequency interval (Hz)	PSD (g^2/Hz)
20–100	3 dB/oct.
100–600	0.2
600	−3 dB/oct.
G_{rms}	16.63 g

In an equivalent static condition, the fixed-free beam and tip mass are loaded under the influence of the gravitational field of 1 g = 9.81 m/s^2. The enforced random acceleration PSD is given in Table 3.1. The objective of the random response analysis is the calculation of the r.m.s. values of the shear force D_A and bending moment M_A. Two analyses are performed:

1. The static equivalent enforced random acceleration method.
2. A complete finite element analysis (FEA).

In both cases is the modal damping ratio is $\zeta = 0.05$, and the associated amplification factor is $Q = 10$.

Analysis 1

The static shear force D_A and the bending moment M_A can be calculated using the equilibrium equations and are, respectively,

$$D_A = g(m_b L_b + m), \tag{3.4}$$

and

$$M_A = g\left(\frac{1}{2}m_b L_b^2 + m L_b\right). \tag{3.5}$$

The natural frequency f_n (Hz) of the fixed-free beam with a tip mass is given in [3]

$$f_n = \frac{1}{2\pi}\sqrt{\frac{3EI}{L_b^3(m + 0.23 m_b L_b)}}, \tag{3.6}$$

but if the tip mass $m = 0$, the natural frequency of the fixed-free beam can be calculated with [4]

$$f_n = \frac{1.8751^2}{2\pi L_b^2}\sqrt{\frac{EI}{m_b}}. \tag{3.7}$$

The r.m.s. value of the tip mass acceleration $\ddot{x}_{rms}$ can be approximated using Miles' equation

$$\ddot{x}_{rms} = \sqrt{\frac{\pi}{2} f_n Q W_{\ddot{U}}(f_n)}. \tag{3.8}$$

The equivalent static approximation of the r.m.s. values of D_A and M_A is given by

$$D_{A,rms} = (m_b L_b + m)\Big|_{1g}\sqrt{\frac{\pi}{2} f_n Q W_{\ddot{U}}(f_n)}, \tag{3.9}$$

and

$$M_{A,rms} = \left(\frac{1}{2} m_b L_b^2 + m L_b\right)\Big|_{1g}\sqrt{\frac{\pi}{2} f_n Q W_{\ddot{U}}(f_n)}. \tag{3.10}$$

The used physical properties and subsequent analysis results are listed in Table 3.2.

Analysis 2

The second dynamic analysis of the fixed-free beam with tip mass will be done using the finite element analysis (FEA) approach. The beam is modeled into ten beam elements with the following mass and stiffness matrix [5]:

Table 3.2 Physical properties and analysis results—Analysis 1

Property	Value	
E (Pa)	70.0×10^9	70.0×10^9
I (m^4)	1.0×10^{-5}	1.0×10^{-5}
L_b (m)	1.25	1.25
m_b (kg/m)	0.5	0.5
m (kg)	10	0
f_n (Hz)	51.93	423.75
$W_{\ddot{U}}(f_n)$ (g^2/Hz)	0.104	0.20
$\ddot{x}_{rms}$ (g)	9.19	36.48
$D_{A,rms}$ (N)	957.76	223.71
$M_{A,rms}$ (Nm)	1,155.0	139.82

$$[M^e] = \frac{mL}{420}\begin{bmatrix} 156 & 22L & 54 & -13L \\ 22L & 4L^2 & 13L & -3L^2 \\ 54 & 13L & 156 & -22L \\ -13L & -3L^2 & -22L & 4L^2 \end{bmatrix}, \tag{3.11}$$

$$[K^e] = \frac{EI}{L^3}\begin{bmatrix} 12 & 6L & 12 & 6L \\ 6L & 4L^2 & -6L & 2L^2 \\ -12 & -6L & 12 & -6L \\ 6L & 2L^2 & -6L & 4L^2 \end{bmatrix}, \tag{3.12}$$

The tip mass is modeled as a point mass. The orthogonal damping matrix $[C]$ can be calculated by the following expression

$$[C] = 2\zeta \sum_{i=1}^{n} \omega_i [M]\{\Phi_i\}\{\Phi_i\}^T [M], \tag{3.13}$$

where ω_i is the natural frequency, $\{\Phi_i\}$ the associated mode shape, and $[M]$ the total mass matrix of the dynamic system. n is the number of analysis degrees of freedom.

The properties were mentioned in Table 3.2. The results of the FEA are presented in Table 3.3.

The PSDs of the acceleration of the tip mass X, the bending moment M_A, and the shear force D_A are presented in Fig. 3.2.

When the modal effective mass $M_{eff}(f_n)$ is very significant with regard to the other modes, the approximation using the equivalent static acceleration field representation in conjunction with Miles' equation is very satisfactory. If the dynamic continuous system has no discrete masses, the approximation using the equivalent static representation is less satisfactory.

Table 3.3 Analysis results—Analysis 2

Analysis results	Value	
M_{total} (kg)	10.625	0.625
f_n (Hz)	51.81	423.75
$M_{eff}(f_n)$ (kg)	10.32 (97.15%)	0.38 (61.27%)
$W_{\ddot{U}}(f_n)$ (g^2/Hz)	0.104	0.20
$\ddot{X}_{rms}$ (g)	9.14	58.19
$D_{A,rms}$ (N)	912.58	142.36
$M_{A,rms}$ (Nm)	1,131.8	124.34

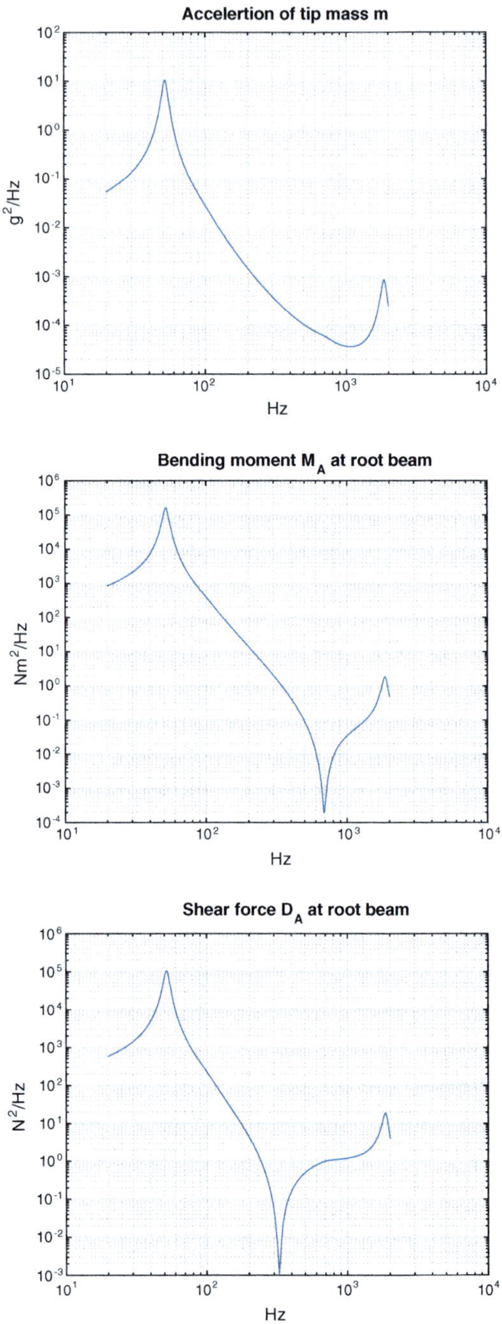

Fig. 3.2 PSD of acceleration tip mass, bending moment M_A, and shear force D_A

3.4 Equivalent Static Force Field

If the SDOF in Fig. 2.6 is loaded by a static force F_{stat}, the static displacement becomes $x_{stat} = F_{stat}/k = F_{stat}/m(2\pi f_n)^2$, or $(2\pi f_n)^4 = (x_{stat}/F_{stat})^2$. Equation (3.3) can be rewritten

$$x_{rms} = \left|\frac{x_{stat}}{F_{stat}}\right| \sqrt{\frac{\pi}{2} f_n Q W_F(f_n)}, \tag{3.14}$$

assuming that the deflected shape caused by the static load F_{stat} is quite similar to the vibration mode $\Phi(f_n)$. The static load F_{stat} may be any static load, e.g., pressure load. In general, we can calculate with the finite element analysis technique the displacement and stresses at arbitrary locations applying unit loads, e.g., $F_{stat} = 1,\ p = 1$. Afterward we multiply the static results by the Miles' equation.

The 'equivalent static force field' method will be illustrated with the example given in Fig. 3.3.

The fixed-free beam has a bending stiffness EI (Nm^2), mass per unit of length $m_b = \rho A$ (kg/m), and length L_b (m). The density of the applied material is ρ (kg/m^3), and the area of the cross section is A (m^2). The material properties and dimensions are given in Table 3.4.

The first natural frequency f_n of the fixed-free beam is given by Eq. 3.7. The associated vibration mode is quite similar to the deflection shape of the beam loaded by a constant line load pb (N/m) as shown in Fig. 3.3. The static bending moment M_A at root of the beam (point A) can be calculated with

$$M_A = \frac{1}{2} p b L_b^2, \tag{3.15}$$

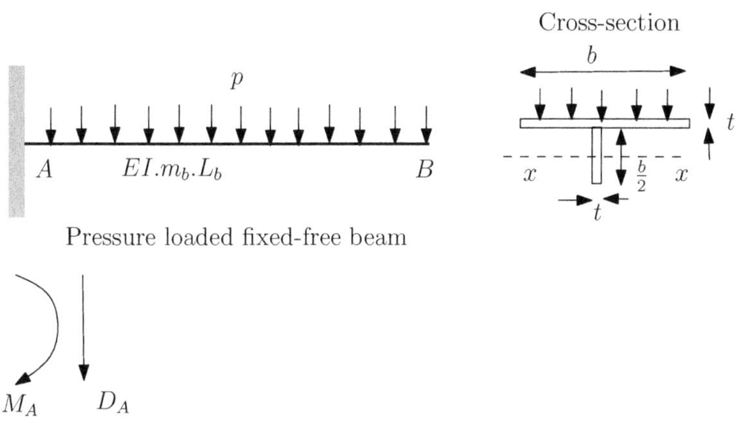

Fig. 3.3 Fixed-free beam pressure loaded

Table 3.4 Material properties and dimension of fixed-free beam

Item	Material property
E (Pa)	70.0^9
ρ (kg/m^3)	2,700
ν (-)	0.33
	Dimension
L_b (m)	1.25
b (m)	0.25
t (m)	0.0005
A (m^2)	$1.5bt$
I (m^4)	$0.14b^3t$

and the shear force D_A is

$$D_A = pbL_b. \tag{3.16}$$

The static equivalent r.m.s. values of the M_A and D_A can be calculated with the following expressions

$$M_{A,rms} = \frac{|M_A|}{p}\sqrt{\frac{\pi}{2} f_n Q W_p(f_n)}, \tag{3.17}$$

and

$$D_{A,rms} = \frac{|D_A|}{p}\sqrt{\frac{\pi}{2} f_n Q W_p(f_n)}, \tag{3.18}$$

where Q is the amplification factor and $W_p(f_n)$ the PSD of the sound pressure at the natural frequency f_n. The characteristics of the sound pressures are given in Table 3.5 [6]. The PSD $W_p(f)$ can be calculated as follows

$$W_p(f) = p_{rms}^2/\Delta f, \tag{3.19}$$

and the mean square of the sound pressure is given by

$$p_{rms}^2 = p_{ref}^2 10^{\frac{SPL}{10}}, \tag{3.20}$$

where $p_{ref} = 2^{-5}$ Pa and the SPL (dB) is the sound pressure level and is given by the following expression

$$SPL = 10 log \frac{p_{rms}^2}{p_{ref}^2}. \tag{3.21}$$

A full finite element analysis will be performed to compute the random dynamic responses of the fixed-free beam exposed to a random dynamic plane wave sound pressure P. The beam is divided into ten beam elements. The mass and stiffness matrix are already given in Eqs. 3.11 and 3.12. The load vector of the beam element

Table 3.5 One octave band sound pressure levels

f_{center} One octave band (Hz)	SPL (dB) $p_{ref} = 2.0^{-5}$ (Pa)	f_{low} (Hz)	f_{upper} (Hz)	Δf (Hz)	$W_p(f_{center})$ Pa^2/Hz
32.5	130	22.3	44.5	22.3	179.58
63	135.5	44.5	89.1	44.5	318.59
125	139	88.4	176.8	88.4	359.47
250	143	176.8	353.6	176.8	451.48
500	138	353.6	707.1	353.6	71.38
1000	132.	707.1	1,414.2	707.1	8.97
2000	128	1,414.2	2,828.4	1,414.2	1.78

loaded by a constant pressure p is taken from [7] and is given by the following expression

$$(F^e) = \frac{pb}{12}\begin{pmatrix} 6L \\ L^2 \\ 6L \\ -L^2 \end{pmatrix}, \tag{3.22}$$

where L is the length of the beam element.

The PSD function of the sound pressure $W_p(f_{center})$ is calculated in Table 3.5 and illustrated in Fig. 3.4. Below the frequency $f = 22.3$, the PSD function $W_p = 0$, because it is outside the SPL frequency interval.

The calculated dynamic characteristics such as undamped natural frequencies, associated modal effective masses, and modal static deflection are presented in Table 3.6.

The modal static deflection is defined by [1]

$$\Gamma_i(f) = \frac{\phi_i^T F_p}{\omega_i^2 \phi_i^T [M]\phi_i}, \quad i = 1, 2, \ldots, n, \tag{3.23}$$

where n is the number of analysis degrees of freedom. The first vibration mode will contribute very significant to the displacement of the fixed-free beam under dynamic random sound pressures.

To perform the dynamic random response analysis, the orthogonal damping matrix (Eq. 3.13) is generated based on the constant modal damping ratio $\zeta = 0.05$ $(Q = 10)$.

The results of the static equivalent random force analysis using Miles' equation and the FEA are presented in Table 3.7.

The quasi-static approximation underestimates the r.m.s. acceleration of the tip of the fixed-free beam; however, the r.m.s. values of the shear force D_A and M_A are over-estimated, especially the shear force D_A.

The PSD of the displacement and acceleration of the tip of the beam and the shear force and bending moment at the root of the beam are shown in Figs. 3.5 and 3.6.

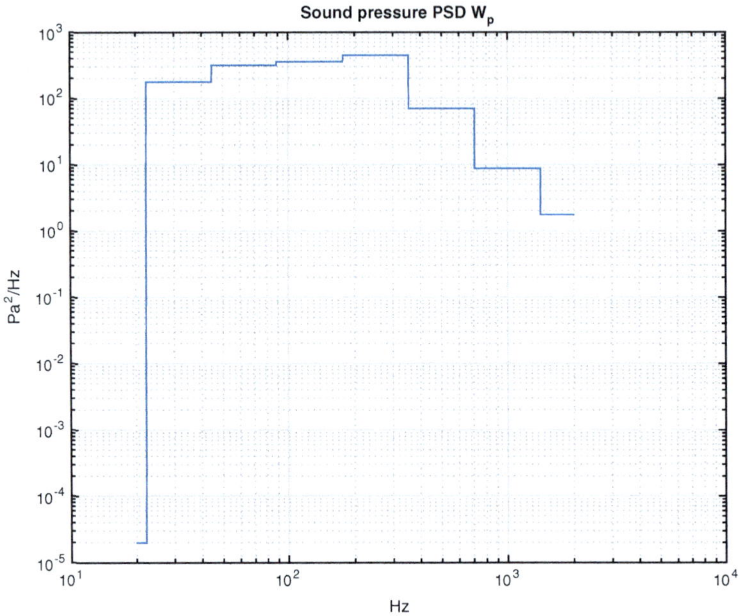

Fig. 3.4 Sound pressure PSD function $W_p(f)$

Table 3.6 Modal characteristics fixed-free beam

Mode ≠	Natural frequencies (Hz)	Modal effective mass (kg)	Modal static deflection ($p = 1$ Pa) $\times 10^6$ (m)
1	139.3	0.39	0.4017
2	872.9	0.12	0.0057
3	2,444.6	0.04	−0.0004

Table 3.7 Comparison analysis results of Miles' equation and FEA

	Analysis results	
Item	Miles's equation ($p = 1$)	FEA ($p = 1$)
X_{rms} (m) tip	2.0^{-3}	2.0^{-3}
$\ddot{X}_{rms}(9)$ tip	154.25	163.61
$F_{A,rms}$ (N)	619.68	384.41
$M_{A,rms}$ (Nm)	387.30	344.78

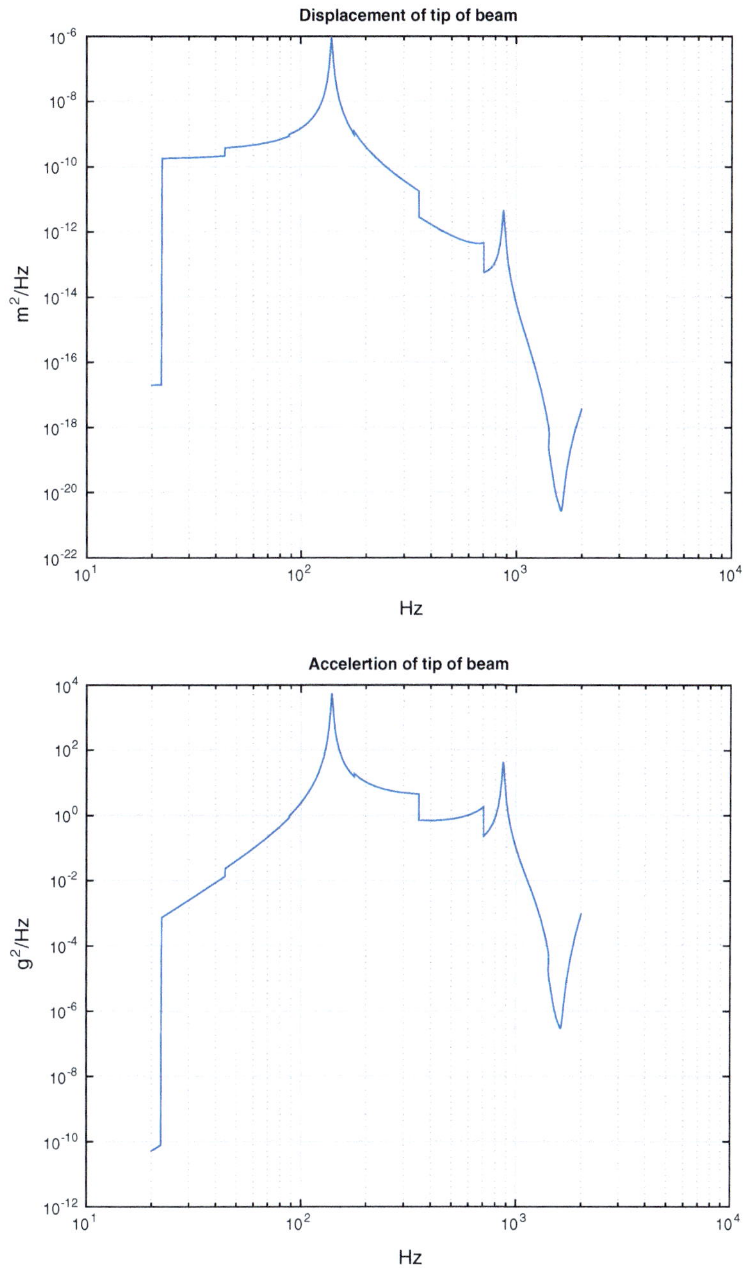

Fig. 3.5 PSD of random responses of displacement and acceleration of the tip of the beam

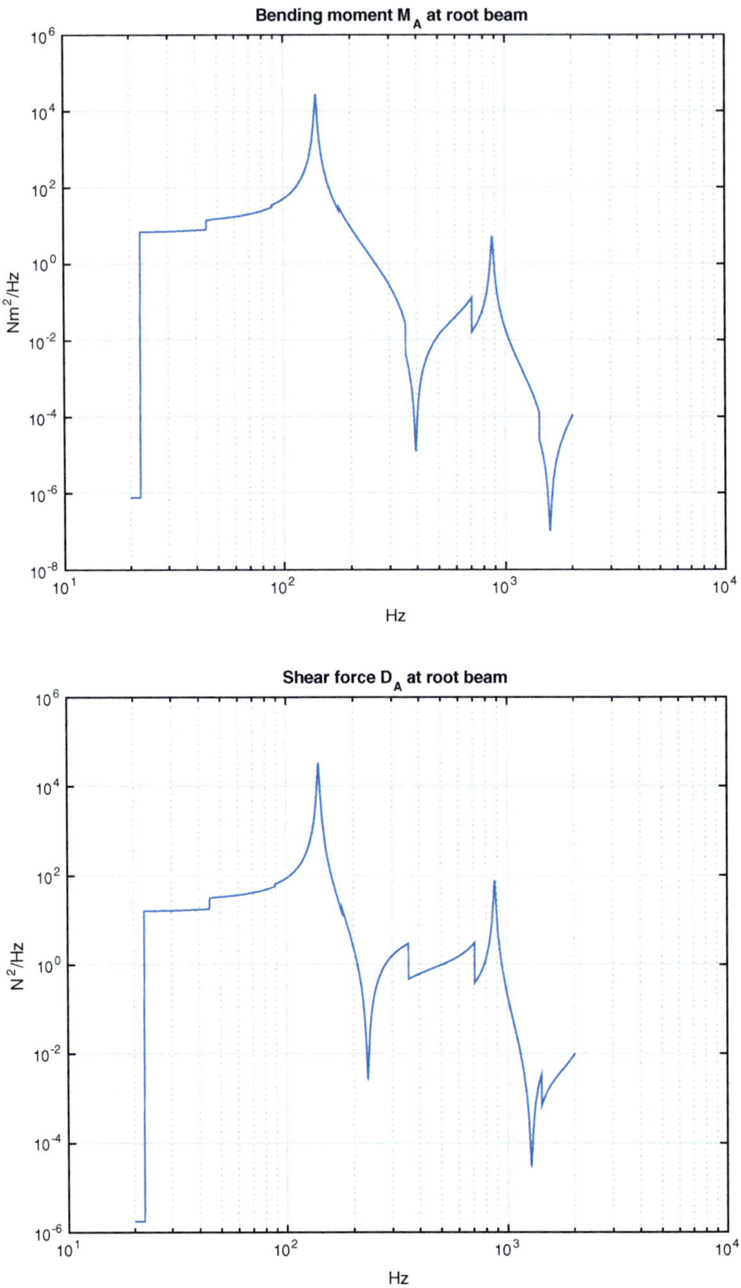

Fig. 3.6 PSD of random responses of the bending moment and shear force at the root of the beam

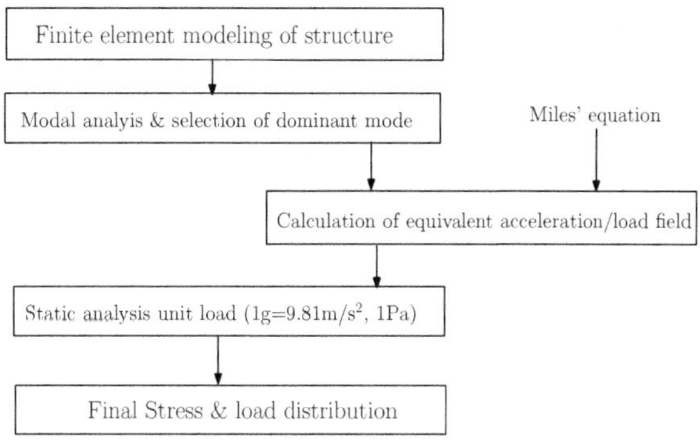

Fig. 3.7 Flowchart equivalent static analysis

3.5 Equivalent Static Finite Element Analysis

In the paper of Daneshjou et al. [2], a flowchart is developed, which shows the procedure to perform equivalent static analysis using the finite element analysis method in combination with Miles' equation. The flowchart is shown in Fig. 3.7.

The subsequent steps in the equivalent static analysis with the finite element analysis method are discussed step by step:

Finite element modeling of the structure The dynamical system (structure, mass distribution, applied materials) is discretized using a mesh of finite elements. In fact, applying the finite element analysis method is the transformation of partial differential equations of motion into matrix equations of motion. Many books to learn about the finite element analysis method are published, e.g., [5, 8, 9]. In summary, the following steps to build a finite element model (FEM) are to be done:

- Build the mesh, place nodes, and assign elements.
- Apply boundary conditions and loads.
- Apply FEM checks; 6×6 mass matrix and location of center of gravity, rigid body zero strain energy, ...

Further some suggestions can be given[1]:

- Simpler models lead to better results.
- Your are more productive with simpler modes.
- There is no way to avoid proper modeling.

Modal analysis and selection of dominant vibration mode In general, the modal analysis is done on an undamped dynamic system, because the influence of the damped natural frequency is not spacecraft structures is not significant. The outcomes of he undamped eigenvalue analysis are:

[1]Lecture notes Dr. Patrick Steffen, Method of Finite Elements, ETH, Zürich.

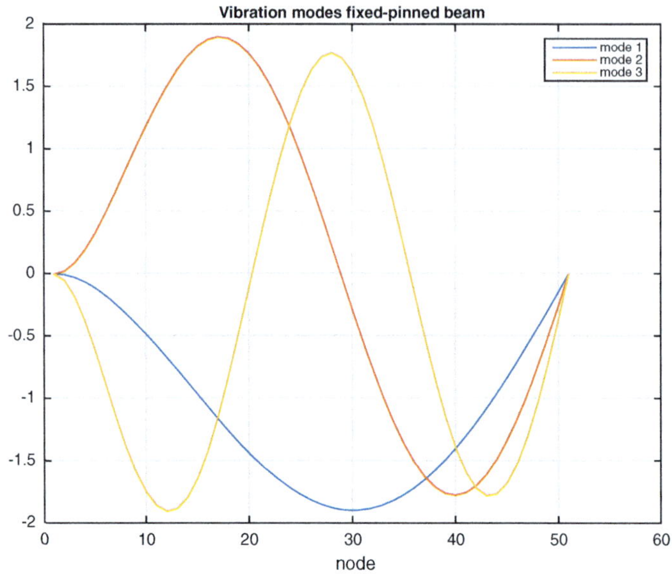

Fig. 3.8 Modal displacements

- The real eigenvalues or undamped natural frequencies.
- The real eigenvectors or real vibration models.
- The modal effective masses [10] in case of enforced acceleration of the system.
- The modal participation factors [1] in case of force loaded systems.

The selection of the dominant mode will be illustrated by the following example. The first three modes are shown in Fig. 3.8, and the associated modal characteristics of the first three vibration modes will be given in Table 3.8

The first vibration mode is dominant when the fixed-pinned beam will be excited by an enforced acceleration or loaded by unitary uniform load (say pressure). The corresponding PSD values of the random enforced acceleration $\ddot{U}$ (m/s^2) and random pressure P(Pa) at the natural frequency of the first mode are given by $W_{\ddot{U}}(f_n)$(g^2/Hz) and $W_p(f_n)$(Pa^2/Hz), respectively.

Calculation of equivalent r.m.s. acceleration/load using Miles' equation Q is the amplification factor and $Q = 1/2\zeta$.

- The equivalent static acceleration can be calculated with Miles' equation; $\ddot{x}_{rms} = \sqrt{\frac{\pi}{2} f_n Q W_{\ddot{U}}(f_n)}$ (g).
- The equivalent static pressure load can be calculated with Miles' equation; $p_{rms} = \sqrt{\frac{\pi}{2} f_n Q W_p(f_n)}$ (Pa).

Static Analysis unitary static load (1 g = 9.81 m/s^2, 1 Pa Perform static FE analysis by applying unitary acceleration (1 g = 9.81 m/s^2) or unitary pressure field (1 Pa). Calculate load and stress distribution in structure at locations of interest.

Table 3.8 Modal characteristics

Mode ≠	Natural frequency f_n (Hz)	Modal effective mass M_{eff} (kg), %	Modal participation factor Γ (−)
1	231	**0.47 (74)**	**−0.34**
2	748	0.00 (1)	0.03
3	1,560	0.07 (11)	−0.13

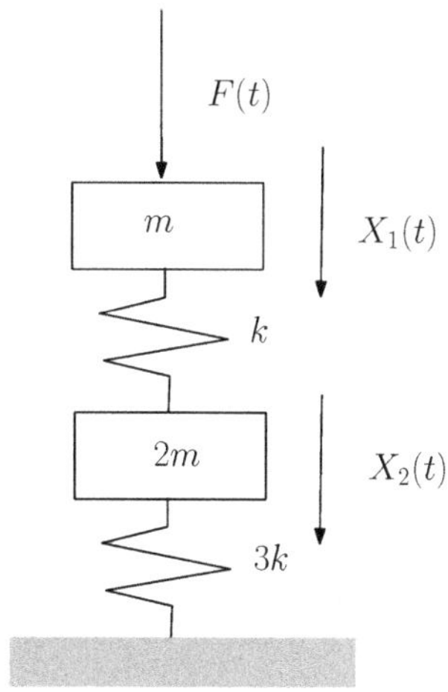

Fig. 3.9 MDOF system loaded by random force

Calculation of load and stress distribution In this final step the load and stress distribution due to the unitary acceleration and pressure loads are multiplied by the appropriate Miles 3σ equivalent static loads to obtain the final load and stress distribution.

Example

The procedure shown in Fig. 3.7 will be illustrated by a simple worked-out example. The MDOF system is shown in Fig. 3.9.

The MDOF system is loaded by a random dynamic force $F(t)$. Calculate the r.m.s. spring force in the spring with spring constant 2 k. We will start to calculate the modal properties (modes, undamped natural frequencies, modal participation factors) of the MDOF system.

- The modal matrix is $[\Phi] = [-0.8881, -0.4597; -0.3251, 0.6280]/\sqrt{m}$.
- The two undamped natural frequencies are $f_n = (0.1267, 0.2448)\sqrt{k/m}$.
- The two modal participation factors are $(\Gamma) = (-0.8881; -0.4597)/\sqrt{m}$.

The first mode with the natural frequency $f_n = 0.1267\sqrt{k/m}$ and associated modal participation factor $\Gamma = -0.8881/\sqrt{m}$ will be selected to be the dominant mode to be applied in conjunction with Miles' equation.

In the static analysis, the dynamic load $F(t)$ is replaced by a unitary force $F_{stat} = 1$ and the displacement vector and spring forces are calculated.

- The static displacement vector associated with the unitary static load $(x_{stat}) = (1.3333; 0.0333)/k$. The ratio of the components of the static displacement vector (x_{stat}) deviates from the ratio of the components of the selected first vibration mode $\Phi(1)$. This may influence the accuracy of the equivalent static approximation.
- The absolute values of the static forces in the springs are $F_k = 1.0000$ and $F_{3k} = 1.0000$.

Calculate the r.m.s. forces in both springs given the modal damping ratio $\zeta = 0.05$, the spring stiffness $k = 10^7$ N/m, $m = 10$ kg, and the white noise PSD of the force $F(t)$ is $W_F = 1000\,\text{N}^2/\text{Hz}$. The undamped natural frequencies are $f_n = 126.72; 244.81$ Hz. The r.m.s forces in the springs can now be obtained

$$(F_{rms}) = \begin{pmatrix} F_k \\ F_{3k} \end{pmatrix} \sqrt{\frac{\pi}{2} f_n Q W_F} = \begin{pmatrix} 1.0000 \\ 1.0000 \end{pmatrix} \sqrt{\frac{\pi}{2} 126.72 \times 10 \times 1000} = \begin{pmatrix} 1410.9 \\ 1410.9 \end{pmatrix} \text{N}.$$

Afterward it is very useful to repeat the calculations of the random vibration problem with your own favorite FEA software package and investigate the accuracy of the static approach in conjunction with Miles' equation.
Answers: $F_{k,rms} = 1187.3$ N and $F_{3k,rms} = 2037.9$ N.

3.6 Chapter Summary

In this chapter, the random dynamic response analysis is replaced by a simple approximate static analysis in combination with Miles' equation. Examples are discussed to illustrate this equivalent static approach.

Problems

3.1 Prove with the aid of Fig. 3.10 that the natural frequency f_n (Hz) of the SDOF system is given by

$$f_n = \frac{1}{2\pi}\sqrt{\frac{1}{\delta_{F=1} m}}. \qquad (3.24)$$

3.2 Calculate the first and second moment of area W_{xx} and I_{xx}, respectively, around xx axis and center of gravity of the T-section in Fig. 3.3, neglecting higher-order terms of the thickness t.
Answers: $0.3333b^2t,\ 0.1389b^3t$

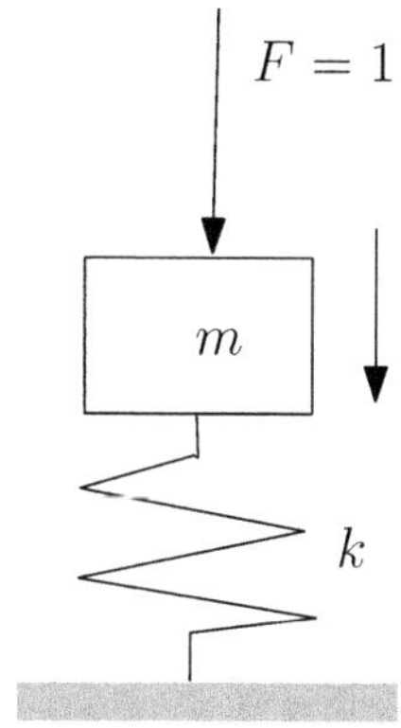

Fig. 3.10 Static displacement of SDOF system

3.3 Calculate the overall SPL of the sound pressures given in Table 3.5.
Answer: OASPL = 146 dB

3.4 The fixed-pinned beam is shown in Fig. 3.3, but in addition the beam is simply supported at point B. The material and geometrical properties are given in Table 3.4, and the PSD function $W_p(f)$ is presented in Table 3.5. The modal damping ratio is $\zeta = 0.05$.

1. Calculate for the fixed-pinned beam the reaction shear forces and bending moment in point A and B; D_A, M_A and D_B, respectively.
2. Calculate with theoretical formula first natural frequency f_n of fixed-pinned beam and associated PSD value $W_p(f_n)$.
3. Calculate with the equivalent force field method $D_{A,rms}$, $M_{A,rms}$, $D_{B,rms}$.
4. Calculate with the FEA[2] the first two natural frequencies fn_i, $i = 1, 2$ and the r.m.s. of the reaction forces and bending moment $D_{A,rms}$, $D_{B,rms}$, and $M_{A,rms}$.

Answers:

1. $D_A = 5/8pbL_b$, $M_A = 1/8pbL_b^2$, and $D_B = 3/8pbL_b$.
2. $f_n = 610.74$ Hz, $W_p(f_n) = 71.38$ Pa²/Hz.
3. $D_{A,rms} = 361.42$ N, $M_{A,rms} = 90.35$ Nm, $D_{B,rms} = 216.85$ N.
4. $f_n = 610.75,\ 1,979.5$ Hz, $D_{A,rms} = 261.53$ N, $D_{B,rms} = 179.27$ N, $M_{A,rms} = 82.36$ Nm.

3.5 A fixed-free (cantilevered) beam with tip mass is illustrated in Fig. 3.11. The bending stiffness of the massless beam is EI (Nm²), and the length is L (m). The tip mass is m (kg), and the area of the massless plate welded to the mass is A_p (m²). The amplification factor is Q. The PSD of the pressure is W_p and is constant in the frequency interval of interest. Calculate the natural frequency f_n (Hz) and the r.m.s. values of the displacement X, the bending moment M_A, and the shear force D_A in accordance with the flowchart given in Fig. 3.7.

[2]Use your favorite FE software package.

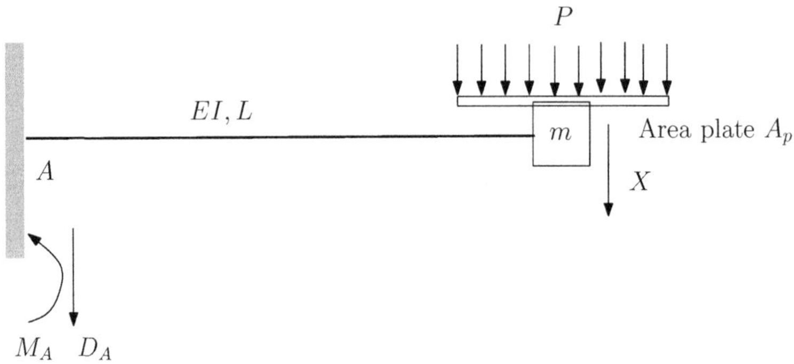

Fig. 3.11 Fixed-free beam with tip mass

Answers: $f_n = \sqrt{3EI/mL^3}$, $x_{rms} = A_p L^3/3EI \times \sqrt{(\pi/2) f_n Q W_p}$, $M_{A,rms} = A_p L \times \sqrt{(\pi/2) f_n Q W_p}$, $D_{A,rms} = A_p \times \sqrt{(\pi/2) f_n Q W_p}$

3.6 For the circular simply supported plate, the radial and tangential bending moments M_r and M_t are calculated when the plate is enforced random accelerated along the rim with a constant single-sided PSD $W_{\ddot{U}}$ (g^2/Hz). The theory of plates and shell and nomenclature are taken from the famous book of Timoshenko and Krieger [11].

The bending stiffness of the plate is given by D, r is the running radius, the outer radius is a, the mass per unit of area is m_p, and Poisson's ratio is ν.

The random dynamic response analyses is performed applying the assumed modes method [12]. There are two different assumed modes considered:

- The static deflection mode $\Phi_1(r) = A(a^2 - r^2)(\frac{5+\nu}{1+\nu}a^2 - r^2)$ taken from [11].
- The assumed mode taken from [13]; $\Phi_2(r) = B(4+\nu)a^3 - 3(2+\nu)ar^2 + 2(1+\nu)r^3$.

Both assumed modes fulfill the boundary condition $\Phi_1(a) = \Phi_2(a) = 0$, and the shapes are similar to the first vibration mode of the plate [4].

Evaluate the natural frequency corresponding to both assumed modes with the aid of Rayleigh quotient.
Answers: $\omega_{n,1}^2 = 24.92D/(ma^4)$, $\omega_{n,2}^2 = 24.89D/(ma^4)$.

The assumed modes shall fulfill the orthogonally relation $\int_A \Phi^2 dA = 1$. Define the scaling constants A and B.
Answers: $A = 0.2120/(\pi a^{10} m)$, $B = 0.1950/(\pi a^8 m)$

Calculate the radial bending moment $M_r(r = 0)$ corresponding to the assumed modes and with unit acceleration $\ddot{u} = 1$ g or unit pressure p = 1 Pa.
Answers: Enforced acceleration: $M_r(r = 0) = -13.32m\ddot{u}\big|_{=1} a^2 AD$, $M_r(r = 0) = -18.59m\ddot{u}\big|_{=1} aBD$, pressure field: $M_r(r = 0) = -13.32p\big|_{=1} a^2 AD$, $M_r(r = 0) = -18.59p\big|_{=1} aBD$

The bending stiffness of the circular sandwich panel is given by $0.5Eh^2t$, where Young's modulus of the face sheets is $E = 70$ GPa, and the thickness of the face

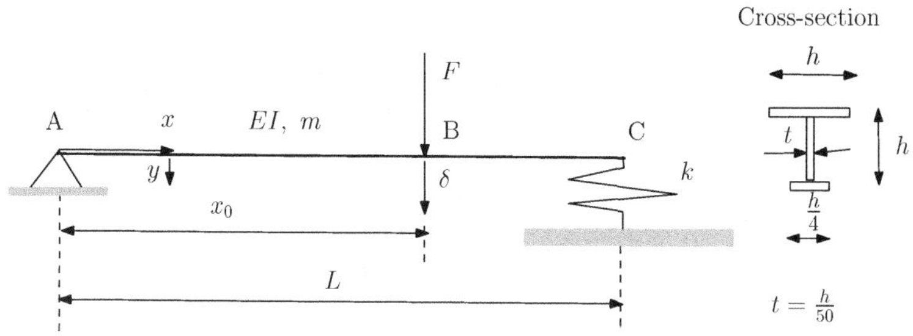

Fig. 3.12 Simply (pinned)-spring supported beam [14]

sheets is $t = 0.5$ mm. The lowest natural frequency of the circular plate is $f_n = 75$ Hz. The diameter of the circular plate is $d = 1$ m. The PSD of the enforced acceleration is $W_{\ddot{U}} = 0.04$ g^2/Hz in between 20–200 Hz. The PSD of the pressure field is $W_p = 100$ Pa^2/Hz in between 20–200 Hz. Define the core height h considering both assumed modes. Calculate for both the enforced random acceleration and random pressure field and both assumed modes the r.m.s. radial bending stress in the face sheets at $r = 0$. The bending resistance is given by $W_b = I/0.5h = ht$.

3.7 A simply spring supported beam is illustrated in Fig. 3.12. The beam is loaded by a random force F with a PSD $W_F(f)$ (N^2/Hz). Calculate the r.m.s. displacement δ_{rms} at location B.

The solution can be achieved by performing the following steps:

- Calculate the static displacement δ for $F = 1$ N using the Myosotis equations [15].
- Calculate with Rayleigh's quotient the approximation of the natural frequency f_n, when the assumed mode is $y(x) = \sin\left(\frac{\pi x}{2L}\right)$.
- Calculate δ_{rms} using Miles' equation Eq. (3.14).

Answers: $\delta|_{F=1} = -\frac{L^3\alpha}{3EI}\left[\alpha^3 - 2\alpha^2 + \alpha(1+\gamma)\right]$, $\alpha = \frac{x_0}{L}$, $\gamma = \frac{3EI}{kL^3}$, $f_n = \frac{\sqrt{32kL^3+\pi^4 EI}}{8\pi\sqrt{m L^2}}$

The second part of this problem is a numerical evaluation:

- The length of the Al-alloy beam is $L = 1$ m, Young's modulus $E = 70$ GPa, the density $\rho = 2,700$ kg/m^3, and the spring stiffness $k = \frac{3EI}{L^3}$ (N/m) ($\gamma = 1$). Define a preliminary cross section (h, t) of the beam such that the lowest natural frequency is $100 \leq f_n \leq 120$ Hz. Use the approximate equation for f_n.
- Repeat the modal analysis with your favorite FEA package and generate a FEM consisting of ten simple (Hermitian) beam elements. Define a more accurate cross section (h, t) to achieve the lowest natural frequency objective.
- Check with your FEM the displacement $\delta|_{F=1}$ for $\alpha = 0.6$.

Answers: $A = \frac{9}{4}ht$, $I \approx \frac{1}{3}h^3 t$, $h = 100$ mm, $t = 2$ mm, $f_n = 108.4461$ Hz, $h = 120$ mm, $t = 2.4$ mm, $f_n = 109.1148$ Hz, $\delta|_{F=1} = 1.4385 \times 10^{-6}$ versus 1.4385×10^{-6} m (FEM)

3.8 Evaluate the strong and weak points of the flowchart of the equivalent static approach in conjunction with Miles' equation as presented in Fig. 3.7.

References

1. Biggs JM (1964) Introduction to structural dynamics. McGraw-Hill Book Company, New York
2. Daneshjou K, Fakoor M (2007) Efficient algorithm for reliability analysis of structures under random vibration. J Solid Mech Mater Eng 1(11):1293–1304
3. Harris CM, Crede CE (1976) Shock and vibration handbook, 2nd edn. McGraw-Hill Book Company, New York. ISBN 0-07-026799-5
4. Blevins RD (1995) Formulas for natural frequency and mode shape. Krieger Publishing Company, New York. ISBN 0-89464-894-2
5. Hwon YW, Bang H (2000) The finite element method using MATLAB, 2nd edn. CRC Press, Boca Raton. ISBN 0-8493-0096-7
6. Wijker JJ (2008) Spacecraft structures. Springer, Berlin. ISBN 978-3-540-755524
7. Kwon YW, Bang H (2000) The finite element method, using MATLAB, 2nd edn. CRC Press, Boca Raton. ISBN 0-8493-0096-7
8. Cook RD (1995) Finite element modeling for stress analysis. Wiley, New York. ISBN 0-471-10774-3
9. Seshu P (2012) Textbook of finite element analysis. PHI Learning. ISBN 978-81-203-2315-5
10. Wijker JJ (2004) Mechanical vibrations in spacecraft design. Springer, Berlin. ISBN 3-540-40530-5
11. Timoshenko S, Woinowsky-Krieger S (1959) Theory of plates and shells. McGraw-Hill Company, Inc., New York
12. Graig RR (1981) Structural dynamics, an introduction to computer methods. Wiley, New York. ISBN 0-471-87715-8
13. Temple G, Bickley WG (1956) Rayleigh's principle and applications to engineering. Dover Publications, New York
14. Kukla S, Owczarck A (2005) Stochastic vibration of a bernoulli-euler beam under random excitation. Sci Res Inst Math Comput Sci 4(1):71–78
15. den Hartog JP (1961) Strength of materials. Dover Publications Inc., New York

Chapter 4
Random Vibration Load Factors

Abstract In general, the structural design of spacecraft structures, instruments, and structural elements is based on the static load factors specified in the launch vehicle (L/V) user's guide(s) or by other quasi-static load (QSL) specifications. In this chapter, the calculated random load factors are based on the 3σ values of the interface loads of a particular system and depend on the random load levels, natural frequencies, and associated modal effective masses. If the random load factors are beyond the specified quasi-static load factors, adaptation (notching) of the random dynamic loads may be considered. Examples and problems with answers included are provided.

Keywords Quasi-static loads · Mass participation · Random load factor · Modal effective mass · 3σ value

4.1 Introduction

In this chapter, the random load factors are derived. These load factors are based on 3σ interface (reaction) forces.

4.2 Three-Sigma Design Approach

The (usually) narrowband random response of a structure to a random excitation yields important information about the durability and strength of the structure. Of great interest is the probability that the response will exceed some critical level. If the response levels in the structure, at the location we are monitoring the response, exceed the allowable strength of the structure at that location, then a failure will occur. Failure may be defined as a yielding of the material causing a permanent deformation or just an excessive deformation causing a foul condition. Other failures may be ductile fracture or buckling under excessive compressive loads. These conditions are called first-passage failure [1] because an undesirable situation exists at the first occurrence

J. Wijker, *Miles' Equation in Random Vibrations*, Solid Mechanics and Its Applications 248, https://doi.org/10.1007/978-3-319-73114-8_4

of the condition; subsequent excursions above the level that caused the initial failure are irrelevant, as the structure has already failed.

We adopt a much simpler (and consequently approximate) approach to the types of failure listed above. The three-sigma (3σ) approach (design) gives a simple design guide to the first-passage problem. If we know the strength of the structure at the point we have monitored the response, employing the three-sigma rule we can say that for a component with a strength of three times the value of the stress in the component due to the random vibration, the stress in the component will exceed its strength only 0.26% of the time assuming a zero mean. Consequently, it is common practice to examine the r.m.s. response quantities from the random response analysis (acceleration, stress, forces, etc.), multiply these values by 3, and use them to compare with the component strengths, e.g., calculating the margin of safety. The best we can do then is to say that, if the component was just strong enough (i.e., its strength was just greater than the stress level $x = 3\sigma$), then there is a 99.74% chance that the component will not suffer a first-passage failure.

4.3 Random Vibration Load Factors

Random vibration load factors are typically calculated from the applicable random vibration criteria using Miles' equation, which is based on statistical analyses of enforced acceleration spectra with a 3σ level. Miles' equation determines a load factor by assuming that the fundamental (first system) mode in each orthogonal direction will provide the primary response:

$$E\{\ddot{X}^2\} = \sigma^2 = \frac{\pi}{2} f_n Q W_{\ddot{U}}(f_n). \tag{4.1}$$

where Q is the amplification factor, f_n is the system fundamental frequency (Hz), $W_{\ddot{U}}(f_n)$ is the enforced acceleration spectral density at f_n (g^2/Hz). For most components, an amplification factor $Q = 10$ for all three directions is common practice if no test data are available. Component frequencies are determined either by finite element modal analysis or sinusoidal sweep test. The $W_{\ddot{U}}(f_n)$ values are determined from the component natural frequency, f_n, and the design random vibration environment, which envelopes the maximum input spectra for a particular mounting location. This method is reasonably accurate for systems with dominant fundamental modes, considering it approximates a component's response using a SDOF system to represent loading over an entire frequency spectrum between 20 and 2000 Hz.

Example

A fixed-free (cantilevered) beam supports at the free end an instrument represented by a SDOF system with mass $M = 10$ kg and spring k (Fig. 4.1).

The lowest natural frequency of that instrument is specified to be $f_n = 125$ Hz. The length of the beam is $L = 1.0$ m. Young's modulus of the Al alloy is $E = 70$ GPa, and the density of the material is $\rho = 2700$ kg/m^3. About the cross section is

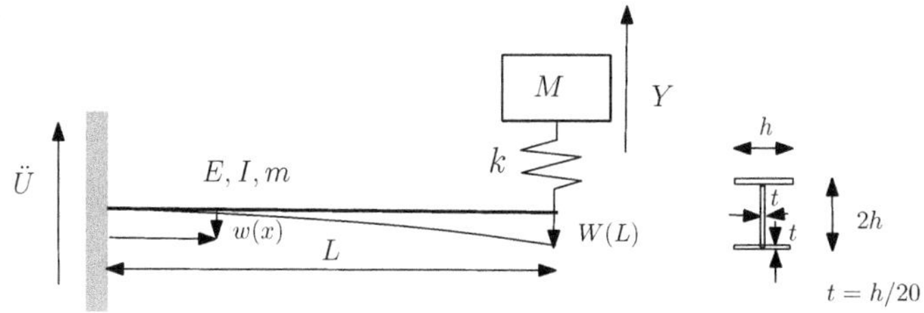

Fig. 4.1 Fixed-free beam

Fig. 4.2 Dunkerley's systems

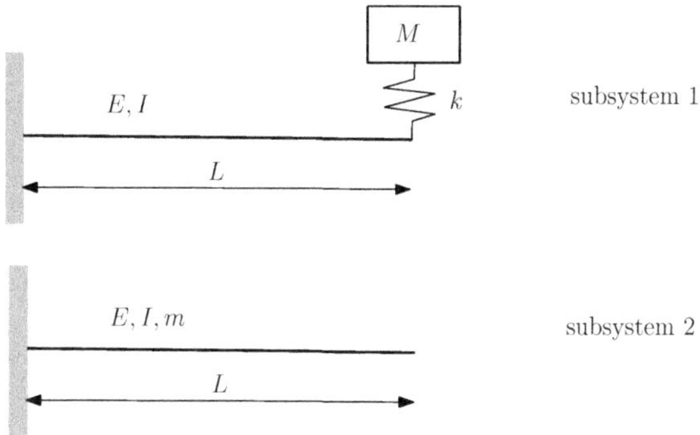

$h = 100$ mm, and the wall thickness of the flanges and central section is $t = h/20$. Estimate the lowest natural frequency of that system applying Dunkerley's method [2], with the knowledge that the lowest natural frequency of the cantilevered beam is $\omega_{cb}^2 = 12.3623EI/mL^4$. To estimate the lowest natural frequency of the dynamic system is divided into two subsystems 1 and 2 as illustrated in Fig. 4.2.

The subsystems have the following natural frequencies:

subsystem 1 $\quad \omega_{11}^2 = (3kEI)/((kL^3 + 3EI)M)$.

subsystem 1 $\quad \omega_{22}^2 = \omega_{cb}^2 = 12.3623EI/mL^4$.

The estimation of the lowest natural frequency of the dynamic system shown in Fig. 4.1 can be calculated by

$$\frac{1}{\omega_{sys,1}^2} = \frac{1}{\omega_{11}^2} + \frac{1}{\omega_{22}^2}. \tag{4.2}$$

The lowest natural frequency is $f_{sys,1} = 67.78\,\text{Hz}$.

The modal damping ratio $\zeta = 0.05$ and the PSD of the random acceleration at $f_{sys,1}$ is $W_{\ddot{U}}(f_{sys,1}) = 0.02\,\text{g}^2/\text{Hz}$. The peak acceleration (inertia load) $\ddot{y}$ is the 3σ value and is calculated by Miles' equation

$$G_{peak} = 3\sqrt{\frac{\pi}{2} f_{sys,1} Q W_{\ddot{U}}(f_{sys,1})} = 13.8\,\text{g}. \tag{4.3}$$

The peak acceleration $\ddot{y}$ will be statically applied to the dynamic system to perform the strength analysis.

Perform the modal analysis with your own favorite finite element analysis (FEA) software package.

4.4 Mass Participation Approach

The mass participation approach had been discussed in [3] to estimate design loads. This method is based on the value of reaction force when the SDOF system is driven by a random enforced acceleration. This is illustrated in Fig. 4.3. Assuming a harmonic enforced acceleration $\ddot{U}(\omega)$, the reaction force $F_{react}(\omega)$ can be expressed as

$$F_{react}(\omega) = m\left[1 + \left(\frac{\omega}{\omega_n}\right)^2 H(\omega)\right]\ddot{U}(\omega), \tag{4.4}$$

where the frequency response function $H(\omega)$ is given by

$$H(\omega) = \frac{1}{1 - \frac{\omega^2}{\omega_n^2} + 2j\zeta\frac{\omega}{\omega_n}}. \tag{4.5}$$

The natural frequency is $\omega_n = 2\pi f_n = \sqrt{k/m}$ and the damping ratio is $\zeta = c/2\sqrt{km}$ and the amplification factor is $Q = 1/2\zeta$. The mean square value of the reaction force $F_{react_{ms}}$

$$F_{react_{rms}} = F^2_{react_{rms}} = \frac{\pi}{2}m^2 f_n Q W_{\ddot{U}}(f_n)(1 + 4\zeta^2) \approx \frac{\pi}{2}m^2 f_n Q W_{\ddot{U}}(f_n). \tag{4.6}$$

For multi-degrees of freedom (MDOF) dynamic system, the mass m must be replaced by the modal effective mass $M_{eff}(f_n)$.

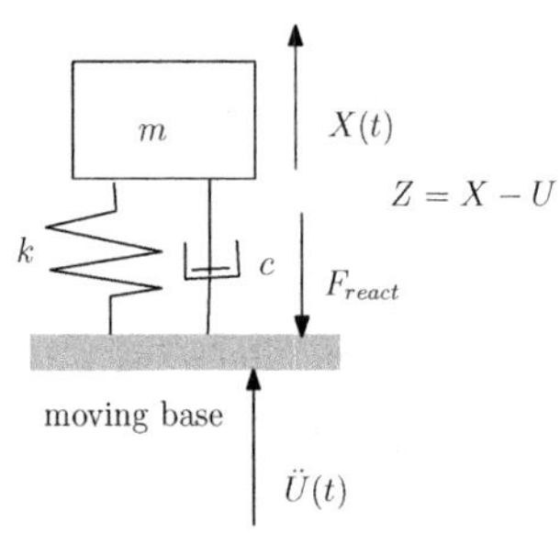

Fig. 4.3 SDOF system

The root sum square (RSS) represents the mean square value of the reaction force all over the frequency band and must be equal the total mass of the MDOF system M_{tot} multiplied by the equivalent acceleration $\ddot{y}$, thus

$$F_{react_{rms}} = \sqrt{\sum_{k=1}^{n} F^2_{react_{rms}}(f_k)} = \sqrt{\sum_{k=1}^{n} \frac{\pi}{2} M^2_{eff}(f_k) f_n Q_k W_{\ddot{U}}(f_k)} = M_{tot}\ddot{y}. \quad (4.7)$$

In accordance with the mass participation approach, the $\ddot{y} = 3\sigma$ (static) design acceleration becomes

$$\ddot{y} = \frac{3}{gM_{tot}} \sqrt{\sum_{k=1}^{n} \frac{\pi}{2} M^2_{eff}(f_k) f_k Q_k W_{\ddot{U}}(f_k)} \text{ (g)}. \quad (4.8)$$

where g is the gravitational constant (9.81 m/s^2). In practice, the total mass is replaced by the summation of the modal effective masses

$$3\ddot{y} = \frac{3}{g\sum_{k=1}^{n} M_{eff}(f_k)} \sqrt{\sum_{k=1}^{n} \frac{\pi}{2} M^2_{eff}(f_k) f_k Q_k W_{\ddot{U}}(f_k)} \text{ (g)}. \quad (4.9)$$

where $\ddot{y}$ is called the 3σ values of the random acceleration (load factors), which can be used for dimensioning of experiments and their interfaces.

The residual mass M_{res} is defined as

$$M_{res} = M_{tot} - \sum_{k=1}^{n} M_{eff}(f_k), \quad (4.10)$$

and, in general, $\sum_{k=1}^{n} M_{eff}(f_k) \geq 0.90 \cdots 0.95 M_{tot}$. Equation (4.8) can be extended to

$$\ddot{y} = \frac{3}{g\sum_{k=1}^{n} M_{eff}(f_k)} \sqrt{\sum_{k=1}^{n} \frac{\pi}{2} M^2_{eff}(f_k) f_k Q_k W_{\ddot{U}}(f_k) + M^2_{res} \int_{20}^{2000} W_{\ddot{U}}(f) df} \text{ (g)}. \quad (4.11)$$

Equation (4.11) can be applied to predict notched random enforced acceleration spectra, e.g.,

$$F_{react_{rms}} = M_{tot}\ddot{y} = 3\sqrt{\sum_{k=1}^{n} \frac{\pi}{2} M^2_{eff}(f_k) f_k Q_k W_{\ddot{U}}(f_k) + M^2_{res} \int_{20}^{2000} W_{\ddot{U}}(f) df} \text{ (N)}. \quad (4.12)$$

Example

A sandwich rectangular plate with a length a and width b is simply supported along the edges. The core of sandwich cross section has a height h, and both Al alloy facings have a thickness t and Young's modulus E. The bending stiffness is D. The mass per unit of area of the plate is m_p. The assumed modes are given by

$$\Phi(x, y)_{m,n} = \sin\left(\frac{mx}{a}\right)\sin\left(\frac{ny}{b}\right). \tag{4.13}$$

The associated natural frequencies can be calculated using

$$f_{m,n} = \frac{\pi}{2}\left[\left(\frac{m}{a}\right)^2 + \left(\frac{n}{b}\right)^2\right]\sqrt{\frac{D}{m_p}},\ n, m = 1, 2, \cdots .. \tag{4.14}$$

and the modal effective mass $M_{eff,m,n}$

$$M_{eff,m,n} = \frac{16\pi^2 m\left(\frac{a}{\pi m} - \frac{a\cos(\pi m)}{\pi m}\right)^2 m_p n\left(\frac{b}{\pi n} - \frac{b\cos(\pi n)}{\pi n}\right)^2}{(a\sin(2\pi m) - 2\pi a m)\ (b\sin(2\pi n) - 2\pi b n)},\ n, m = 1, 2, \cdots . \tag{4.15}$$

The random acceleration vibration specification $W_{\ddot{U}}$ is given in Table 4.1.

The dimensions and properties of the simply supported rectangular plate are given in Table 4.2

The analysis results of the load factor $\ddot{\gamma}$ are plotted in Fig. 4.4. The load factor slowly decreases when the number of participating vibration modes is increasing. The load factor $\ddot{\gamma} = 58.8$ g with an one first mode approximation with a natural frequency $f_n = 119.66$ Hz. The modal effective mass $M_{eff,m,n} = 0$ when the wave number m, n are even.

Example

A scientific instrument can be simply represented by a SDOF system with a mass $m = 20$ kg and a natural frequency $f_n = 125$ Hz. The damping ratio is $\zeta = 5\%$. The quasi-static design load factor is 15 g. The PSD value of random enforced acceleration is $W_{\ddot{U}}(125) = 0.05\ \text{g}^2/\text{Hz}$. The 3σ acceleration $\ddot{\gamma}$ can be calculated with the aid of Miles' equation

$$\ddot{\gamma} = 3\sqrt{\frac{\pi}{2} f_n Q W_{\ddot{U}}(125)} = 29.7\ \text{g}.$$

Table 4.1 Random acceleration vibration specification

Frequency (Hz)	PSD (g^2/Hz)
20–100	3 dB/oct.
100–600	0.4
600–2000	−3 dB/oct.
G_{rms}	22.54 g

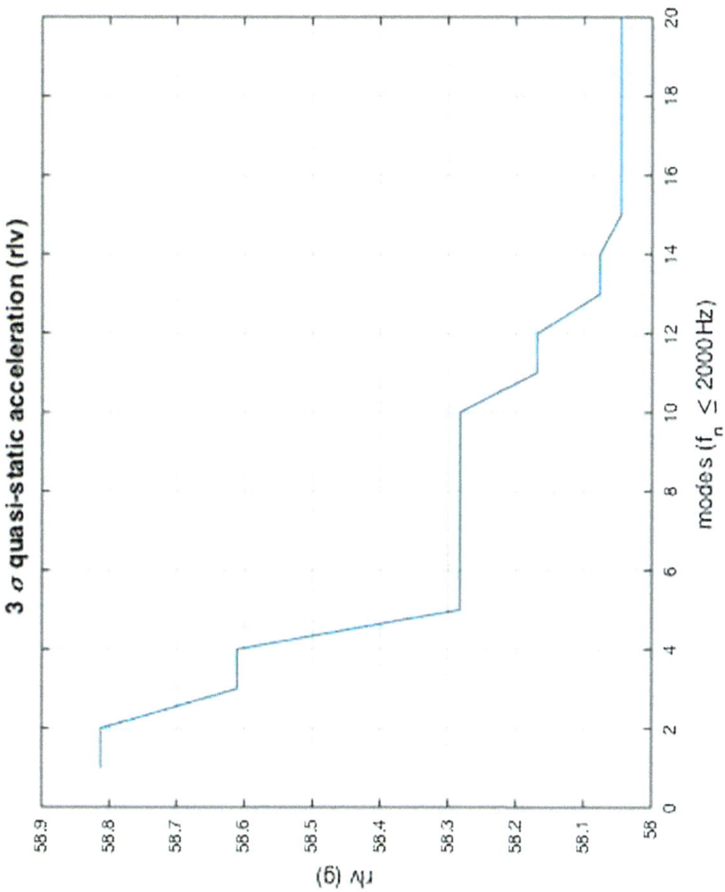

Fig. 4.4 Random load factor $\ddot{y}$ (rlv) dependent on number of modes ($f_n \leq 2000$ Hz)

Table 4.2 Dimensions and properties rectangular plate

Item	*Notation	Unit	Value
Young's modulus	E	Pa	7.0×10^{10}
Length	a	m	0.5
Width	b	m	0.4
Face thickness	t	m	1.0×10^{-4}
Core height	h	m	3.0×10^{-3}
Density facings	ρ	kg/m^3	2700
Mass per unit area	m_p	kg/m^2	$2\rho t + 0.5$
Second moment of area	I	m^3	$0.5h^2 t$
Bending stiffness	D	Nm	EI
Damping	ζ	–	0.05

The PSD of the enforced random acceleration shall be modified (notched) such to meet the quasi-static design load factor

$$W^*_{\ddot{U}}(125) = \frac{15^2}{3^2 \frac{\pi}{2} f_n Q} = 0.013 \text{ g}^2/\text{Hz}.$$

4.5 Chapter Summary

In general, the design of spacecraft, instrument, and structural elements is based on the static load factors provided in the L/V user's guide or by other quasi-static load (QSL) specifications. The random load factor $\ddot{\gamma}$ is based on the 3σ value of the interface loads. If the random load factor is beyond the static load factor, reduction (notching) in the random dynamic loads shall be considered.

Problems

4.1 The uniform cantilevered beam with added mass and stiffness is shown in Fig. 4.5.

The assumed mode of the cantilevered beam is $\Phi(x) = x^2(6L^3 - 4Lx + x^2)$. The physical displacement can be written as $w(x) = \eta(t)\Phi(x)$.

- Calculate the lowest natural frequency ω_n^2 with the aid of Lagrange's equation.
- Calculate the lowest natural frequency ω_n^2 with the aid of Dunkerley's method (beam + SDOF).

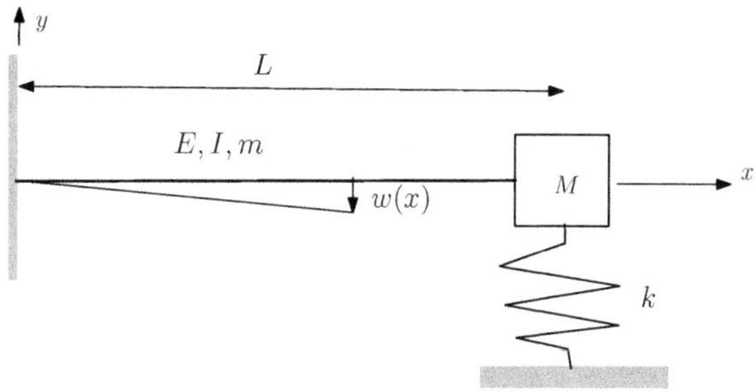

Fig. 4.5 Sketch of the beam with added mass and stiffness [4]

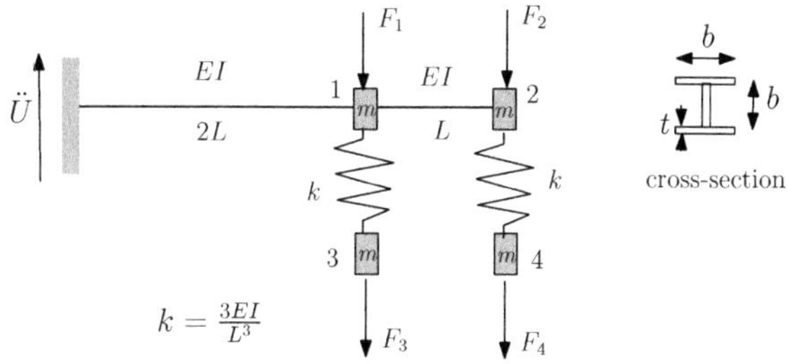

Fig. 4.6 Collatz beam problem [5]

answers: $\omega_n^2 = (9kL^8 + 28.8EIL^5)/(9ML^8 + 2.311mL^9)$, $\omega_n^2 = (162kEI)/(162EIM + 13kmL^4)$.

4.2 A massless cantilevered beam with discrete masses is shown in Fig. 4.6. This example is taken from a classical book about eigenvalue problems [5]. This example is selected to practice the Myosotis method [6] of cantilevered beams. The beam is loaded at the root with an enforced random acceleration $\ddot{U}$, which is specified in Table 4.1. The bending stiffness of the beam is EI and the total length $3L$. The four masses m are equal. The stiffness of the spring is $k = 3EI/L^2$.

The force vector is given by $\{F\} = \{F_1, F_2, F_3, F_4\}^T$ and the displacement vector $\{X\} = \{X_1, X_2, X_3, X_4\}^T$ and the components are in the same direction as the forces. The relation between force vector and displacement vector is the flexibility matrix $[G]$, such that $[G]\{F\} = \{X\}$. The numerical values of the parameters are: $E = 7 \times 10^{10}$ Pa, $b = 30$ cm, $t = 1$ mm, $m = 5$ kg, $L = 1/3$ m, $\zeta = 0.05$, $I = (7/12)b^3t$ m^4.

- Setup the flexibility matrix $[G]$.
- Calculate natural frequencies $\omega_n^2 = (2\pi f_n)^2$.
- Calculate natural frequencies f_n and associated modal effective masses M_{eff} using numerical values of the parameters.
- Calculate random load factor $\ddot{\gamma}$ taking into account:
  - First vibration mode.
  - All vibration modes.

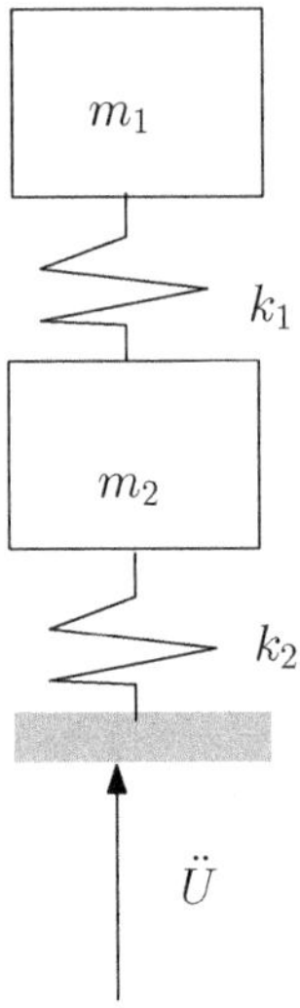

Fig. 4.7 Two SDOF systems

Answers: $[G] = \frac{L^3}{3EI}\begin{pmatrix} 8 & 14 & 8 & 14 \\ 14 & 27 & 14 & 27 \\ 8 & 14 & 9 & 14 \\ 14 & 27 & 14 & 28 \end{pmatrix}$, $\omega_n^2 = (0.0433, 1.6237, 6.0439, 9.5392)$ EI/mL^3 Hz,
$f_n = 80.77, 494.8, 954.7, 1199$ Hz, $M_{eff} = 18.27, 1.52, 0.00, 0.21$ kg, $\ddot{\gamma} = 55.81, 56.97$ g.

4.3 Two SDOF systems are shown in Fig. 4.7. The masses are $m_1 = 10$ kg and $m_2 = 25$ kg, respectively. The spring stiffnesses are $k_1 = 2.0 \times 10^6$ N/m and $k_2 = 1.0 \times 10^7$ N/m. The two SDOF systems are loaded by a random enforced acceleration at the base, which are specified in Table 4.1. The modal damping ratio is $\zeta = 0.04$.

- Calculate natural frequencies f_n and associated modal effective masses M_{eff}.
- Calculate random load factor $\ddot{\gamma}$ taking into account:
  - The first vibration mode.
  - Both vibration modes.

Answers: $fn = 61.91, 115.72$ Hz, $M_{eff} = 22.54, 12.47$ kg, $\ddot{\gamma} = 38.46, 41.58$ g.

4.4 A fixed-supported beam with length $L_b = 2$ m, Young's modulus $E = 70$ GPa, bending stiffness EI, and mass per unit length $m_b = \rho A + 0.5$ kg/m is shown in Fig. 4.8. In the cross section is $b = 0.4$ m and the thickness $t = 0.5$ mm. The density of the metal is $\rho = 2700$ kg/m^3 and the modal damping $\zeta = 0.05$. The box has a mass $M = 20$ kg and is placed in the middle of the beam.

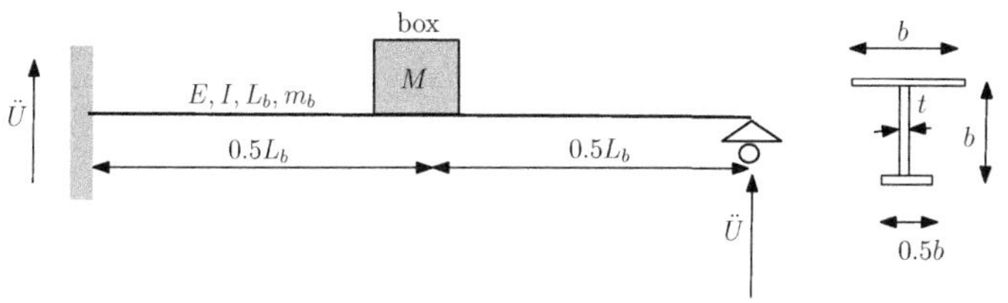

Fig. 4.8 Fixed-supported beam

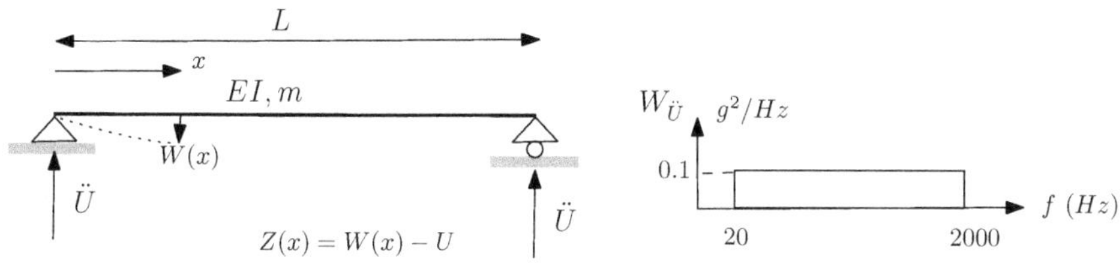

Fig. 4.9 Simply supported beam

- Calculate area A and second moment of area I.
- Calculate total mass M_{tot}.
- Generate a finite element model.
- Calculate in between 20 and 2000 Hz the natural frequencies f_n and associated modal effective masses M_{eff}. Calculate the residual mass M_{res}
- Calculate the load factor $\ddot{\gamma}$ when the beam is excited by a random acceleration as given in Table 4.1.

Answers: $A = 2.5bt$, $I = 0.267b^3t$, $M_{tot} = 23.70$ kg, $f_n = 97.95, 1048.2, 1826.9$ Hz, $M_{eff} = 22.59, 0.01, 0.52$ kg, $M_{res} = 0.57$ kg, $\ddot{\gamma} = 70.38$ g.

4.5 Shown in Fig. 4.9 is a simply supported beam excited at the supports by an enforced random acceleration $\ddot{U}$. This problem is discussed at the Web site www.caeai.com of CAE Associates. Derive the expression for the r.m.s. relative displacement $Z(x)$ as given in Eq. (4.16) using the simply supported beam. The r.m.s. relative displacement $z_{rms}(x)$ can be calculated using the following expression

$$w_{rms}(x) \approx z_{rms}(x) = \sqrt{\frac{\pi \Gamma^2 \Phi(x)^2 W_{\ddot{U}}(f_n)}{4\zeta m_g^2 (2\pi)^4 f_n^3}}, \tag{4.16}$$

with the following parameters:

- The PSD of the enforced acceleration at f_n, $W_{\ddot{U}}(f_n)$.
- fundamental natural frequency f_n (Hz) associated with vibration mode $\Phi(x) = \sin(\pi x/L)$, (derive $f_n = 1.5708\sqrt{\frac{EI}{mL^4}}$).

- modal participation factor $\Gamma = -m \int_0^L \Phi(x)dx$.
- generalized mass $m_g = m \int_0^L \Phi^2(x)dx$.
- modal damping ratio ζ.

The undamped equation of motion of the beam for the relative displacement $Z(x)$ is given by

$$EIZ^{IV}(x) + m\ddot{Z}(x) = -m\ddot{U}, \tag{4.17}$$

where EI is the bending stiffness, m is the constant mass per unit of length, and $\ddot{U}$ is the random enforced acceleration. The absolute displacement is the summation of the relative displacement and the enforce displacement, $W(x) = Z(x) + U$. The damping is proportional to the velocity $\dot{Z}(x)$ and may be introduced later on. Equation (4.17) can be used to derive Eq. (4.16)

The length of the beam is $L = 2540\,\text{mm}$, Young's modulus $E = 210\,\text{GPa}$, the density $\rho = 7800\,\text{kg/m}^3$, the width $b = 25.4\,\text{mm}$, and the thickness $h = 2.54\,\text{mm}$. Calculate the following parameters:

- the fundamental frequency $f_n = 92.6313\,\text{Hz}$.
- the generalized mass $m_g = 0.06391$.
- the modal participation factor $\Gamma = 0.08137$.
- the modal effective mass $M_{eff} = \Gamma^2/m_g = 0.1036\,\text{kg}$ (prove expression).
- damping ratio $\zeta = 0.02$.
- $\Phi(0.5L) = 1$.
- the PSD of $W_{\ddot{U}}(f_n) = 9.6236\,\text{m/s}^2$.
- the r.m.s. relative displacement $z_{rms}(0.5L) = 7.0325 \times 10^{-04}\,\text{m}$.

References

1. Wirsching PH, Paez TL, Ortiz H (1995) Random vibrations, theory and practice. Wiley, New York. ISBN 0-471-58579-3
2. Wijker JJ (2008) Spacecraft structures. Springer, Berlin. ISBN 978-3-540-755524
3. Chung YT, Krebs DJ, Peebles JH (2001) Estimation of payload random vibration loads for proper structure design. In: 42nd AIAA/ASME/ASCE/AHS/ASC structures, structural dynamics, and materials conference, AIAA-2001-1667, Seattle, 16–19 Apr 2001
4. Stephen NG (1982) Note on the combined use of dubkerley's and southwell's methods. J Sound Vib 83(4):585–587
5. Collatz L (1949) Eigenwertaufgaben. Akademische Verlaggesellschaft Geest & Portig K.-G, Leipzig
6. den Hartog JP (1961) Strength of materials. Dover Publications Inc., Mineola

Chapter 5
Notching and Mass Participation

Abstract The mass participation is one of the approaches for notching analysis, in which the modal effective mass and apparent mass in conjunction with Miles' equation are the basic elements to determine notched random acceleration input comparing 3σ reaction loads with the reaction loads caused by the quasi-static design loads (QSL). The quasi-static design limit load applied for the design of equipment and instruments is mostly based on experience from previous spacecraft projects and is defined, in general, using the mass acceleration curve (MAC). Procedures to calculate random load factors and the estimation of the depth of the notch (modification of input spectrum) using the mass participation are discussed. Examples and problems with answers are given.

Keywords Modal effective mass · Mass participation · Mass acceleration curve · Notching · 3σ value

The mass participation is a approach for notching analysis, in which the modal effective mass and apparent mass in conjunction with Miles' equation are the basic ingredients to determine notched random acceleration input comparing 3σ reaction loads with the reaction loads caused by the quasi-static design loads (QSL).

The quasi–static design limit load applied for the design of equipment and instruments is mostly based on experience from previous spacecraft projects and is defined, in general, using the mass acceleration curve (MAC). Such a curve can be derived from analytical and flight data and includes the effects of both transient and mechanically transmitted random vibration [1, 2]. An example of a MAC is presented In Fig. 5.1.

The design loads (factors) are dependent on the mass of the equipment or the instrument. The design factor (inertia load) $\ddot{\gamma}_{stat}$ (g) defines the interface load between the equipment, the instruments, and the spacecraft to which the boxes are mounted. The maximum random interface loads will occur approximately at the natural frequencies. If the random interface load is higher than the interface loads caused by the design limit load, the input acceleration levels may be reduced (notched) at natural frequencies.

An instrument is represented by a SDOF system enforced by a random acceleration $\ddot{U}$ represented by the PSD function $W_{\ddot{U}}(f)$, which is illustrated in Fig. 5.2. The 3σ

J. Wijker, *Miles' Equation in Random Vibrations*, Solid Mechanics and Its Applications 248, https://doi.org/10.1007/978-3-319-73114-8_5

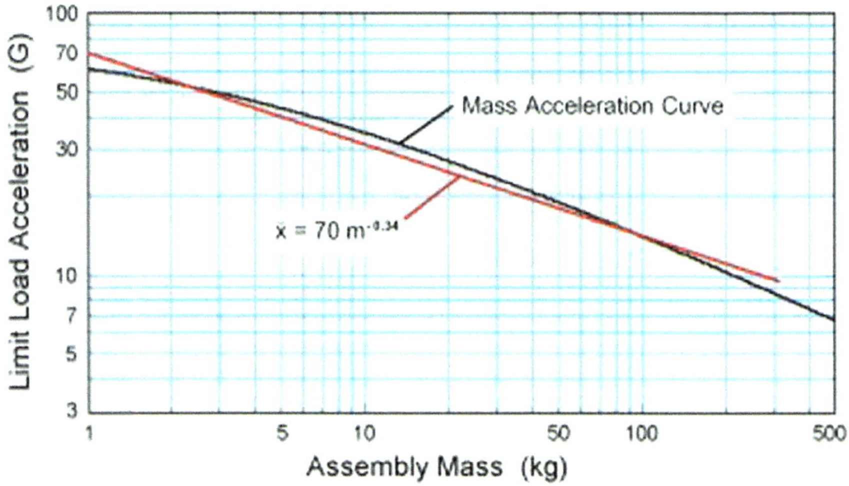

Fig. 5.1 Representative mass acceleration curve for cryogenic assemblies [3]

Fig. 5.2 SDOF system excited by random acceleration

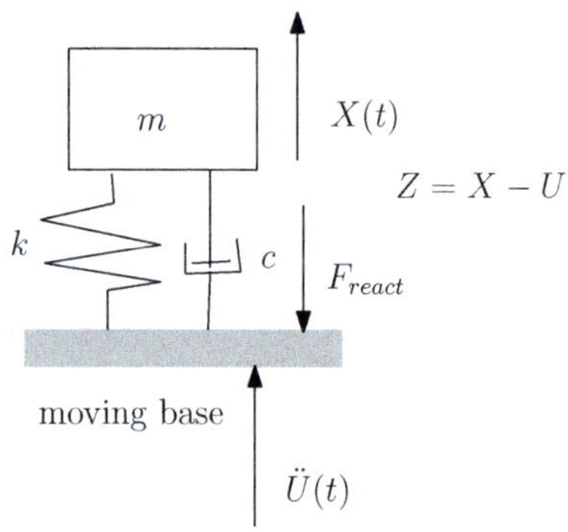

value of the random base force F_{base} with a PSD $W_{F_{base}}(f)$ will be less or equal to the total inertia force $m\ddot{\gamma}_{stat}$ applying at the CoG of the SDOF system. Applying Miles' equation [4] will result in

$$3\sqrt{\int_{f_{min}}^{f_{max}} W_{F_{base}}(f)df} = 3m\ddot{X}_{rms} \approx 3m\sqrt{\frac{\pi}{2} f_n Q W_{\ddot{U}}(f_n)} \leq m\ddot{\gamma}_{stat}, \tag{5.1}$$

where $W_{F_{base}}(f)$ is the PSD function of the random interface loads, f_n(Hz) is the natural frequency, Q is the amplification factor of the SDOF system, and m is the mass. f_{min} is the minimum frequency of interest and f_{max} the maximum frequency of interest, most times 20 and 2000 Hz, respectively. $\ddot{\gamma}_{stat}$ is the load factor. The probability the random acceleration is beyond or below the $\pm 3\sigma$ value is given by

$\text{Prob}(|X| >= 3\sigma) = 0.997$. It is common practice in the European space industry to take 3σ values for comparison with QSL.

In general, the notch criteria with regard to the quasi-static design load (QSL) is defined as

$$3\sqrt{\int_{f_{min}}^{f_{max}} W_{F_{base}}(f)df} \leq M_{tot}\ddot{\gamma}_{stat}, \tag{5.2}$$

where M_{tot} the total mass of the equipment or instrument. Equation (5.2) means that the 3σ ($3 \times rms$ and zero mean) value of the interface force will be less than or equal to the interface load caused by the QSL design loads.

The expected 3σ peak value of the interface force F_{base} is based on the mass participation method

$$F_{base} = 3\sqrt{\sum_{k=1}^{n}\left[M_{eff}^2(f_k)\frac{\pi}{2}f_k Q_k W_{\ddot{u}}(f_k)\right]} \leq M_{total}\ddot{\gamma}_{stat}, \tag{5.3}$$

where n is the number of modes taken into account. The sum of all modal effective masses $M_{eff}(f_k),\ i = 1, 2, \ldots, n$ is equal to the total mass M_{tot} of the system. If m ($m < n$) modes are considered, the effect of the residual mass M_{res} will be taken into account

$$M_{res} = M_{tot} - \sum_{k=1}^{m} M_{eff}(f_k). \tag{5.4}$$

Equation (5.3) will be as follows:

$$F_{base} = 3\sqrt{\sum_{k=1}^{m}\left[M_{eff}^2(f_k)\frac{\pi}{2}f_k Q_k W_{\ddot{u}}^*(f_k)\right] + M_{res}^2\int_{f_{min}}^{f_{max}} W_{\ddot{u}}^*(f)df} \leq M_{tot}\ddot{\gamma}_{stat}, \tag{5.5}$$

where $W_{\ddot{u}}^*(f_k)$ and $W_{\ddot{u}}^*(f)$ are the adapted enforced random acceleration spectra, $W_{\ddot{u}}(f_k)$ the original specified random acceleration spectrum, and $f_k,\ k = 1, 2, \ldots, m$ the natural frequencies. The random enforced acceleration spectrum may be adapted (notched) when the 3σ random interface loads are higher than the quasi-static interface loads.

The modal power contribution of the vibration mode with natural frequency f_k will be related to the original PSD value of the enforced random acceleration $W_{\ddot{u}}(f_k)$. The modal power is defined by

$$M_{eff}(f_k)\ddot{x}_{rms}(f_k)\dot{x}_{rms}(f_k) = M_{eff}(f_k)\dot{x}_{rms}(f_k)\sqrt{\frac{\pi}{2} f_k Q_k W_{\ddot{u}}(f_k)},\ k = 1, 2, \ldots, n. \tag{5.6}$$

The modal power contribution is now correlated to the $W_{\ddot{u}}(f_k)$ as follows

$$M_{eff}(f_k)\ddot{x}_{rms}(f_k)\dot{x}_{rms}(f_k) = A W_{\ddot{u}}(f_k),\ k = 1, 2, \ldots, n, \tag{5.7}$$

where $\ddot{x}_{rms}(f_k)$ is the r.m.s. acceleration, $\dot{x}_{rms}(f_k)$ is the r.m.s. velocity of the vibration mode, A is a constant, and $W_{\ddot{u}}(f_k)$ is the original PSD function of the enforced acceleration at the natural frequency f_k. Rewrite Eq. (5.7) as follows

$$M_{eff}(f_k)\frac{\ddot{x}^2_{rms}(f_k)}{2\pi f_k} = A W_{\ddot{u}}(f_k). \tag{5.8}$$

By introducing of Miles' equation for the r.m.s. acceleration $\ddot{x}(t)$, the following expression can be obtained:

$$M_{eff}(f_k)\frac{\frac{\pi}{2} f_k Q_k W^*_{\ddot{u}}(f_k)}{2\pi f_k} = \frac{M_{eff}(f_k)}{4} Q_k W^*_{\ddot{u}}(f_k) = A W_{\ddot{u}}(f_k). \tag{5.9}$$

The constant A is assumed to be equal for all n vibration modes in the frequency range of interest and can be written as

$$\frac{M_{eff}(f_k) Q_k W^*_{\ddot{u}}(f_k)}{W_{\ddot{u}}(f_k)} = 4A = \bar{A} \quad k = 1, 2, \ldots, m. \tag{5.10}$$

If all vibration modes are considered, the residual mass $M_{res} = 0$ and Eq. (5.5) becomes

$$3\sqrt{\sum_{k=1}^{n}\left[M^2_{eff}(f_k)\frac{\pi}{2} f_k Q_k W^*_{\ddot{u}}(f_k)\right]} \le M_{tot}\ddot{\gamma}_{stat}. \tag{5.11}$$

After substituting Eq. (5.10) into Eq. (5.11), the following expression will be obtained:

$$3\sqrt{\frac{\pi}{2}\bar{A}}\sqrt{\sum_{k=1}^{n}\left[M_{eff}(f_k) f_k W_{\ddot{u}}(f_k)\right]} \le M_{tot}\ddot{\gamma}_{stat}. \tag{5.12}$$

Using Eq. (5.12), the constant $\bar{A}$ can be solved

$$\bar{A} = \frac{(M_{tot}\ddot{\gamma}_{stat})^2}{\frac{9\pi}{2}\sum_{k=1}^{n}\left[M_{eff}(f_k) f_k W_{\ddot{u}}(f_k)\right]}. \tag{5.13}$$

With use of Eq. (5.10), all n individual reduced input spectra $W^*_{\ddot{u}}(f_k),\ k = 1, 2, \ldots, n$ can be calculated.

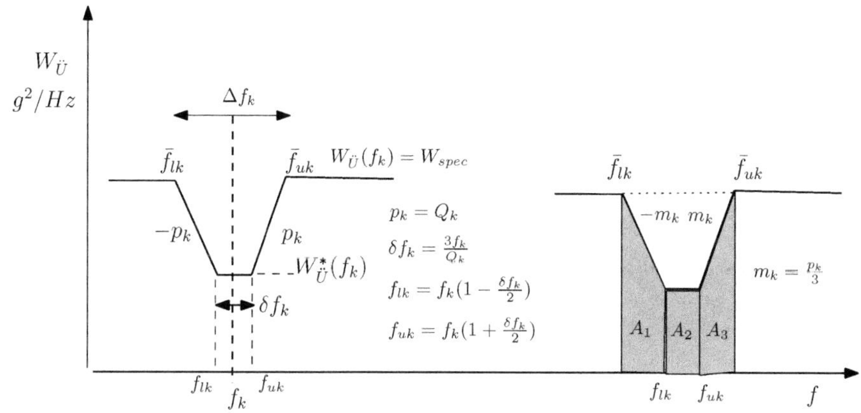

Fig. 5.3 Shape of the notch

If the residual mass is not negligible (say $M_{res} > 0.2M_{total}$), Eq. (5.5) must be applied to obtain the reduced PSD of $W^*_{\ddot{u}}(f_k) \leq W_{\ddot{u}}(f_k)$ at the natural frequencies f_k. This is an iterative process.

The choice of width of the notch around f_k (if needed) is not always chosen to be three times the half power width which depends on the amplification factor Q_k:

$$\Delta f_k = \frac{3f_k}{Q_k} \quad k = 1, 2, \ldots, m, \tag{5.14}$$

and the slopes p_k of the reduced PSD function $W^*_{\ddot{u}}$ around f_k are may be chosen as follows

$$p_k = \mp Q_k \text{ dB/oct.} \tag{5.15}$$

This shape of the notch is illustrated in Fig. 5.3. The natural frequencies and associated amplification factors may be extracted from a low-level sine sweep test with a slow sweep rate.

It is important to know the reduction in G_{rms} when notches are applied. The area A_{notch} can be calculated as follows

$$A_{notch} = \Delta f_k W_{\ddot{U}}(f_k) - A_1 - A_2 - A_3, \tag{5.16}$$

where the areas $A_1.A_2, A_3$ are shown in Fig. 5.3.

At first, the frequencies $f_{uk}, f_{lk}, \bar{f}_{uk}, \bar{f}_{lk}$ have to be calculated. The frequencies f_{lk}, f_{uk} are given by

$$\begin{aligned} f_{lk} &= f_k - \frac{\delta f_k}{2}, \\ f_{uk} &= f_k + \frac{\delta f_k}{2}, \end{aligned} \tag{5.17}$$

and the frequencies $\bar{f}_{lk}$, $\bar{f}_{uk}$ are given by

$$\bar{f}_{lk} = f_{lk}\left(\frac{W^*_{\ddot{U}}(f_k)}{W_{\ddot{U}}(f_k)}\right)^{\frac{1}{m_k}},$$
$$\bar{f}_{uk} = f_{uk}\left(\frac{W_{\ddot{U}}(f_k)}{W^*_{\ddot{U}}(f_k)}\right)^{\frac{1}{m_k}}, \tag{5.18}$$

where $m_k = \frac{p_k}{3} = \frac{Q_k}{3}$.

Now the areas A_1, A_2, A_3 will be calculated using the Eqs. provided in Appendix B.

$$A_1 = \frac{W_{\ddot{U}}(f_k)}{1 - m_k}\left[\left(\frac{f_{lk}}{\bar{f}_{lk}}\right)^{1-m_k} - 1\right],$$
$$A_2 = \delta f_k W^*_{\ddot{U}}(f_k), \tag{5.19}$$
$$A_3 = \frac{W_{\ddot{U}}(f_k)}{1 + m_k}\left[1 - \left(\frac{f_{uk}}{\bar{f}_{uk}}\right)^{1+m_k}\right].$$

Applying the symbolic solver software packages wxMaxima™ the normalized area of the notch $\frac{A_{notch}}{f_k W_{\ddot{U}}(f_k)}$ with $\alpha = \frac{W^*_{\ddot{U}}(f_k)}{W_{\ddot{U}}(f_k)} \leq 1$ for particular values of the amplification factor $Q_k = 5, 10, 15, 20$ can be obtained:

$$\left.\frac{A_{notch}}{f_k W_{\ddot{U}}(f_k)}\right|_{Q_k=5} = -\frac{-36\,\alpha^{\frac{11}{5}} + 140\,\alpha^{\frac{9}{5}} - 39\,\alpha - 65\,\alpha^{\frac{3}{5}}}{80\,\alpha^{\frac{6}{5}}},$$
$$\left.\frac{A_{notch}}{f_k W_{\ddot{U}}(f_k)}\right|_{Q_k=10} = -\frac{-117\,\alpha^{\frac{8}{5}} - 483\,\alpha + 2210\,\alpha^{\frac{9}{10}} - 1610\,\alpha^{\frac{3}{10}}}{1820\,\alpha^{\frac{3}{5}}},$$
$$\left.\frac{A_{notch}}{f_k W_{\ddot{U}}(f_k)}\right|_{Q_k=15} = -\frac{-3\,\alpha^{\frac{7}{5}} - 22\,\alpha + 135\,\alpha^{\frac{3}{5}} - 110\,\alpha^{\frac{1}{5}}}{120\,\alpha^{\frac{2}{5}}}, \tag{5.20}$$
$$\left.\frac{A_{notch}}{f_k W_{\ddot{U}}(f_k)}\right|_{Q_k=20} = -\frac{-207\,\alpha^{\frac{13}{10}} - 2193\,\alpha + 17020\,\alpha^{\frac{9}{20}} - 14620\,\alpha^{\frac{3}{20}}}{15640\,\alpha^{\frac{3}{10}}}.$$

Varying the amplification factor Q_k and the PSD ratio α, the normalized area of the notch $\frac{A_{notch}}{f_k W_{\ddot{U}}(f_k)}$ is shown in Fig. 5.4. When the notches occur at locations on one of the slopes, it is assumed that the normalized notch area is about the same as if the notch is located on the horizontal part of the spectrum as illustrated in Fig. 5.3.

Example

An orbital replacement unit (ORU) had been launched by the space transportation system (STS) and has a mass $m = 187$ kg [5]. The ORU had been designed against a qualification static load factor $\ddot{\gamma}_{stat} = 18.8\,$g, in all three orthogonal directions X, Y, and Z. For a unit with a mass of 187 kg, the qualification random acceleration

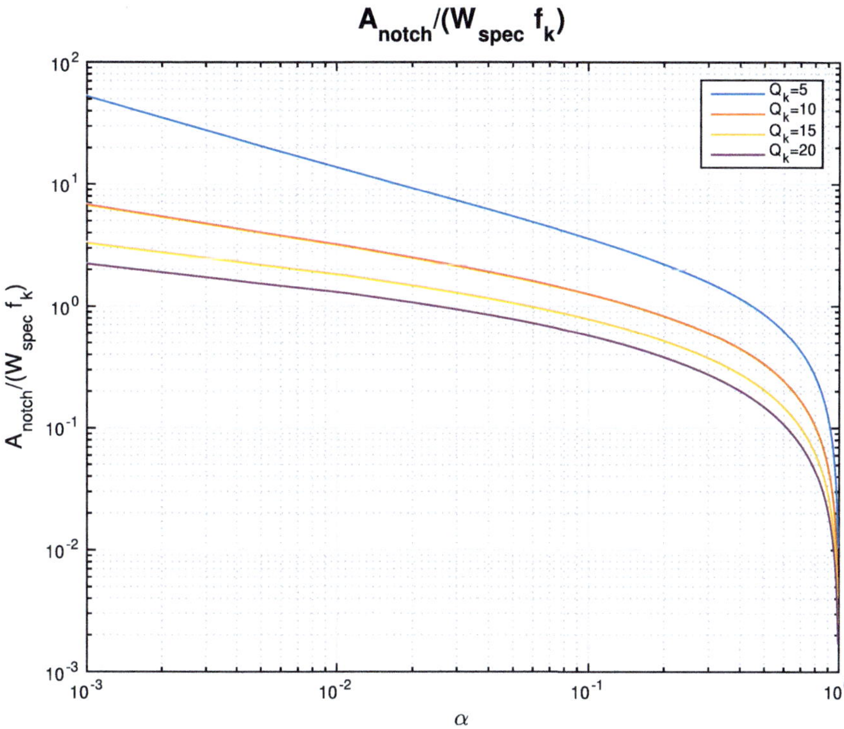

Fig. 5.4 Normalized notch area, $W_{spec} = W_{\ddot{U}}(f_k)$

Table 5.1 Random acceleration vibration specification, qualification level

Frequency (Hz)	PSD (g^2/Hz)
20–80	6 dB/oct.
80–500	0.18
500–2000	−4.5 dB/oct.
G_{rms}	13.02 g

specification is given in the General Environmental Verification Standard (GEVS) [6] and shown in Table 5.1.

The dominant modal effective masses M_{eff} are given in % of the total mass $M_{tot} = 187$ kg and are presented in Table 5.2.

Calculate the random load factors $\ddot{\gamma}$ in X-, Y- and, Z-direction with a modal damping ratio $\zeta = 0.05$ and using the following expression

$$\ddot{\gamma} = \frac{3}{M_{tot}} \sqrt{\sum_{k=1}^{3} \left[M_{eff}^2(f_k) \frac{\pi}{2} f_k Q_k W_{\ddot{u}}(f_k) \right]}. \tag{5.21}$$

Table 5.2 Dominant modal effective masses

		M_{eff}		
Mode	Frequency (Hz)	X %	Y%	Z %
1	59.02	0.395	0.014	31.60
4	74.96	0.119	43.83	0.107
7	92.66	24.50	0.112	16.01

Table 5.3 PSD enfored acceleration $W_{\ddot{U}}(f_k)$

	$f_1 = 59.02$ Hz	$f_2 = 74.96$ Hz	$f_3 = 92.66$ Hz
$W_{\ddot{U}}(f_k)$ (g^2/Hz)	0.0980	0.1580	0.1800

Table 5.4 Random load factors $\ddot{\gamma}$

	X	Y	Z
$\ddot{\gamma}$ (g)	11.8974	17.9349	11.9202

The PSD of the enforced acceleration $W_{\ddot{U}}(f_k)$ at the natural frequencies is provided in Table 5.3.

The random load factors $\ddot{\gamma}$ can be computed and are given in Table 5.4.

All random load factors are $\ddot{\gamma} \leq \ddot{\gamma}_{stat}$; hence no over-testing of the ORU during the random vibration testing will occur.

5.1 Chapter Summary

In this chapter, procedures to calculate random load factors and the estimation of the depth of the notch (modification of input spectrum) with the aid of mass participation using Miles' equation are discussed.

Problems

5.1 A massless simply supported circular plate with bending stiffness D supports at the center a mass-spring system with mass M and spring stiffness k. Poisson's ratio is $\nu = 0.3$. The radius of the circular plate is $r = a$. The system is shown in Fig. 5.5. The simplified system is illustrated in Fig. 5.6. The deflection in the center of the circular plate due to an unit load $F = 1$ is given by [7]

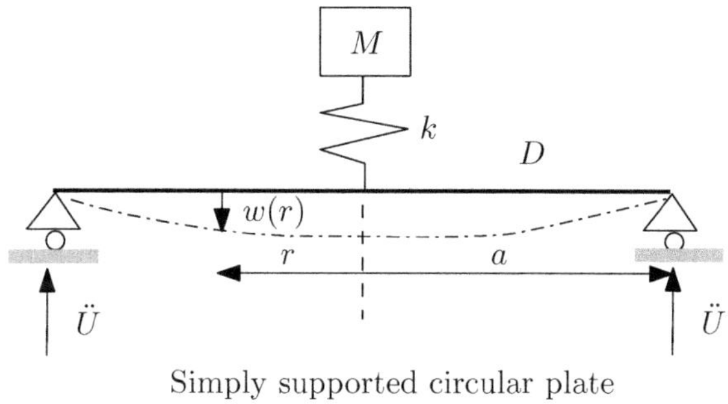

Fig. 5.5 Simply supported circular plate and SDOF element

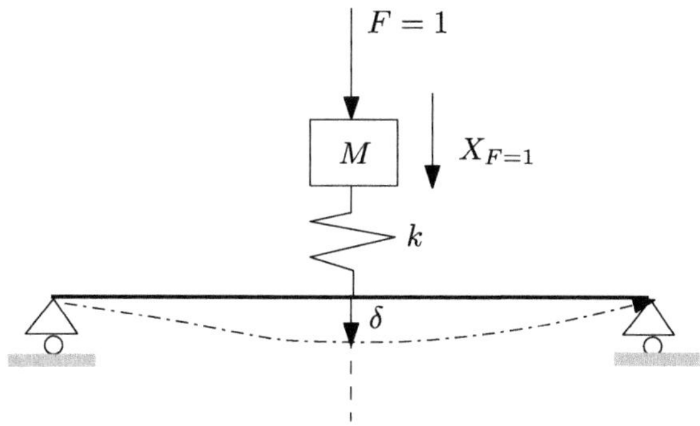

Fig. 5.6 Simplified system

$$\delta = \frac{(3+\nu)a^2}{16\pi(1+\nu)D}. \tag{5.22}$$

- Calculate the lowest natural frequency f_n (Hz) of the complete system, when $k = 1/\delta$.
- The simply supported plate is excited by a random enforced acceleration $\ddot{U}$ (PSD $W_{\ddot{U}}$). The SDOF system had been qualified against $\ddot{y}$ QSL. Describe the procedure to define a notch.

Answers: $f_n = \frac{1}{2\pi}\sqrt{\frac{1}{2\delta M}}$ Hz.

5.2 This problem is the continuation of Problem 5.1. The simply supported plate has a mass per unit of area m_p. Use Dunkerley's method to calculate the lowest natural frequency f_n of the dynamic system presented in Fig. 5.5. The Dunkerley's systems are shown in Fig. 5.7. The deflection $w(r)$ can be written as [8]

$$w(r,t) = \eta(t)\Phi(r) = \eta(t)\left[(4+\nu)a^3 - 3(2+\nu)ar^2 + 2(1+\nu)r^2\right] \tag{5.23}$$

The natural frequency ω_{11}^2 can be obtained using Rayleigh's quotient. The natural frequency ω_{21}^2 can calculated with the displacement method with $F = 1$ ($k = 1/\delta$). Describe the procedure to define a notch when the panel will be overloaded at $r = 0$.

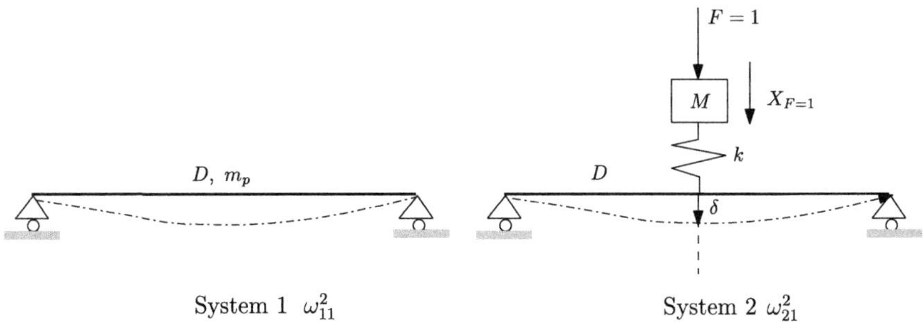

Fig. 5.7 Dunkerley systems

Fig. 5.8 3 SDOF system

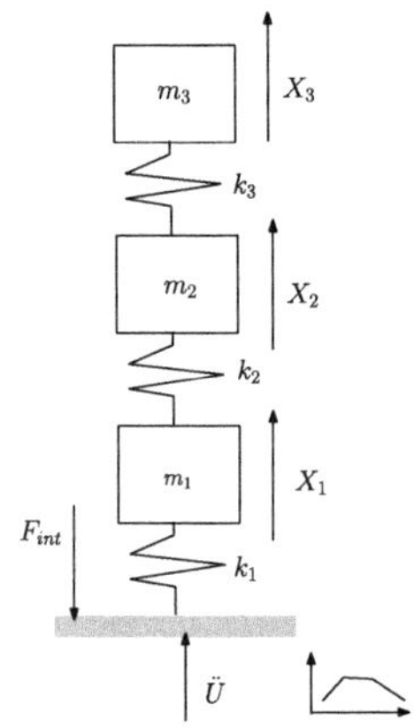

Answers: $1/\omega_n^2 = 1/\omega_{11}^2 + 1/\omega_{21}^2$, $\omega_{11}^2 = 24.4491 D/a^4 m_p$, $\omega_{21}^2 = 9.9008 D/a^2 M$.

5.3 A 3 DOF system is shown in Fig. 5.8. This dynamical system is excited by a random enforced acceleration $\ddot{U}$. The spectrum is given in Table 5.1. The discrete masses are $m_1 = 200$, $m_2 = 100$, $m_3 = 50$ kg, and the stiffness of the springs are $k_1 = 3 \times 10^8$, $k_2 = 2 \times 10^8$, $k_3 = 1 \times 10^8$ N/m, respectively. The modal damping ratio $\zeta = 0.05$. The static design load factor $\ddot{\gamma}_{stat} = 25$ g. Investigate if a modification (notches) of the random input spectrum is needed.

- Calculate the natural frequencies f_k and associated modal effective masses M_{eff}.
- Calculate random load factor $\ddot{\gamma} = 3F_{rms}/M_{tot}$(g) (Eq. 5.3).
- Define $W_{\ddot{U}}(f_k)$ (g).
- Calculate the factor $\bar{A}$.
- Calculate $W_{\ddot{U}}^*(f_k) \leq W_{\ddot{U}}(f_k)$ (g).
- Define A_{notch} (g^2).
- Calculate G_{rms} of modified input spectrum.
- Calculate new random load factor $\ddot{\gamma}$.

Answers: $f_k = 119.95, 239.34, 343.95\,\text{Hz}$, $M_{eff} = 297.56, 47.30, 5.14\,\text{kg}$, $\ddot{\gamma} = 48.16\,\text{g}$, $W_{\ddot{U}}(f_k) = 0.18, 0.18, 0.18\,\text{g}$, $\bar{A} = 616.77$, $W^*_{\ddot{U}}(f_k) = 0.0373, 0.18, 0.18\,\text{g}$, $A_{notch} = 17.61, 0, 0\,\text{g}^2$, $G_{rms} = 12.33\,\text{g}$, $\ddot{\gamma} = 23.88\,\text{g}$.

References

1. JPL. Nasa preferred reliability practices, combination methods for deriving structural design loads considering vibro–acoustic etc., responses. NASA Preferred Reliability Practices Practice No. PD-ED-1211, Jet Propulsion Laboratory (JPL)
2. Trubert M (1989) Mass acceleration curve for spacecraft structural design. Report JPL D-5882, NASA, Jet Propulsion Laboratory
3. Roberts JB, Spanos PD (2003) Random vibration and statistical linearization. Dover Publications Inc., New York. ISBN 0-486-43240-8
4. Wijker JJ (2009) Random vibrations in spacecraft structures design, theory and applications. Number SMIA 165 in solid mechanics and its applications. Springer, Berlin. ISBN 978-90-481-2727-6
5. Fitzpatrick K, McNeill SI (2007) Methods to specify random vibration acceleration environments that comply with force limit specifications. In: IMAC-XXV: conference & exposition on structural dynamics-smart structures and transducers
6. GSFC (ed.) (2013) General environmental verification standard (GEVS), volume GSFC-STD-7000A of GSFC technical standards program. NASA Goddard Space Flight Center, Greenbelt, Maryland 20771, 22 Apr 2013
7. Timoshenko S, Woinowsky-Krieger S (1959) Theory of plates and shells. McGraw-Hill Company, Inc., New York
8. Temple G, Bickley WG (1956) Rayleigh's principle and applications to engineering. Dover Publications, Mineola

Chapter 6
Acoustic and Random Vibration Test Tailoring

Abstract When developing a qualification test program for spacecraft, it is necessary to determine whether there should be an acoustic or random vibration test for each instrument, component, etc. The decision (key) factor is the area/mass ratio of an instrument, component, etc., which is very helpful to make a choice to perform either a random vibration on a shaker table or an acoustic test in a acoustic reverberant chamber. The calculation of the decision factor is completely based on random response analyses applying Miles' equation.

Keywords Random vibration test · Acoustic test

6.1 Introduction

When developing a qualification test program for spacecraft, it is necessary to determine whether there should be an acoustic or random vibration test for each component (instrument, box, etc.). Several criteria must be considered when making such a decision.

First, frequency sensitivity of the component must be established. While many subsystems are not sensitive to frequencies above 2000 Hz, those that do respond above this threshold cannot be tested by random vibration but must instead be tested with acoustic noise. In these cases, decisions must be made regarding precisely how to conduct the acoustic test. The problem is easy for high area-to-mass components, which can be simply suspended with bungee cords within a reverberant acoustic chamber and directly 'hit' with the acoustic source.

Low area-to-mass components are somewhat more problematic, as the vibration energy input should come through the base or mounting structure in the same manner as random vibration. This is because when incident acoustic energy excites a spacecraft structure, high- frequency random vibration is passed through the structure to mounted components.

To simulate this in an acoustic test, the component should be mounted on a plate. The area distribution and dynamic properties of the plate must be similar to the spacecraft structure, so that a 'flight-like' vibration environment is imparted to the

J. Wijker, *Miles' Equation in Random Vibrations*, Solid Mechanics
and Its Applications 248, https://doi.org/10.1007/978-3-319-73114-8_6

unit under test. The plate should then be excited by acoustic energy, which will be passed along to the component in the form of high-frequency random vibration.

Below 2000 Hz, low area/mass components can be tested with random vibration only. The random vibration input specification should be a combination of acoustic and random response of the spacecraft structure at the component interface.

The 'break even' area/mass ratio, representing the regime in which both acoustic and random vibration tests induce equal stochastic acceleration responses, can be calculated analytically for a SDOF system. A schematic of an example component is shown in Fig. 6.1. The example component is a SDOF system having a natural frequency f_n (Hz), a mass m, and a modal damping ζ. From the two simple SDOF systems, the 'break even' area/mass ratio can be calculated. The area/mass ratio A/m will be derived by applying Lyallpur's equation. The damped equations of motion of both SDOF systems (left and right Fig. 6.1) can be written as follows:

$$\begin{aligned} \ddot{Z}(t) + 2\zeta\omega_n\dot{Z}(t) + \omega_n^2 Z(t) &= -\ddot{U}, \\ \ddot{X}(t) + 2\zeta\omega_n\dot{X}(t) + \omega_n^2 X(t) &= \frac{AP(t)}{m} = \frac{F(t)}{m}, \end{aligned} \tag{6.1}$$

where the relative motion of the left SDOF system is $Z = X - U$, the enforced acceleration is $\ddot{U}(t)$, A is the surface area of the right SDOF, and $P(t)$ the plane wave pressure field exposed to the surface area of the SDOF systems. The natural frequency of both SDOF system is $\omega_n^2 = k/m$, the damping is $c/m = 2\zeta\omega_n$, and

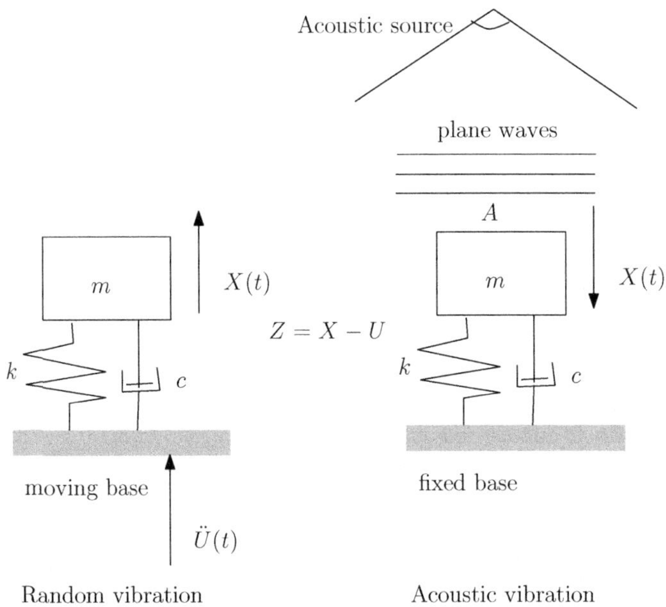

Fig. 6.1 Simple component under acoustic and random vibration [1, 2]

the damping ratio is ζ $(c = 2\zeta\sqrt{km})$. The enforced acceleration and the exposed pressure have a random nature with PSD's $W_{\ddot{U}}(f)$ and $W_P(f)$, respectively. PSD of the equivalent force $F(t)$ is given by $W_{FF}(f)$. The left SDOF in Fig. 6.1 will be denoted by 'Base-Excited SDOF System' and the right SDOF by 'Pressure-Loaded SDOF System'.

The random responses for Z and X for both equations of motion Eq. (6.1) will be found applying Lyapunov's equation.

6.2 Base-Excited SDOF System

The SDOF damped equation of motion with the enforced acceleration $\ddot{U}(t)$ will be expressed in terms of the relative displacement $z(t) = x(t) - u(t)$

$$\ddot{Z}(t) + 2\zeta\omega_n\dot{Z}(t) + \omega_n^2 Z(t) = -\ddot{U}. \tag{6.2}$$

Equation (6.2) will be transformed into space-state variables $Y_1 = Z(t)$ and $Y_2 = \dot{Z}(t)$. The state-space equation is

$$\begin{pmatrix} \dot{Y}_1 \\ \dot{Y}_2 \end{pmatrix} = \begin{pmatrix} 0 & 1 \\ -\omega_n^2 & -2\zeta\omega_n \end{pmatrix}\begin{pmatrix} Y_2 \\ Y_2 \end{pmatrix} + \begin{pmatrix} 0 \\ -1 \end{pmatrix}\ddot{U},$$
$$\{\dot{y}\} = [A]\{Y\} + \{B\}\ddot{U}.$$

The spring force F_k and the force in the damper F_c are considered to be the output variables

$$\{F_o\} = \begin{pmatrix} F_k \\ F_c \end{pmatrix} = \begin{pmatrix} k & 0 \\ 0 & c \end{pmatrix}\begin{pmatrix} Y_1 \\ Y_2 \end{pmatrix} = \begin{pmatrix} \omega_n^2 m & 0 \\ 0 & 2\zeta\omega_n m \end{pmatrix}\begin{pmatrix} Y_1 \\ Y_2 \end{pmatrix} = [C]\{y\}. \tag{6.3}$$

The autocorrelation function of the enforced acceleration $\ddot{U}$ is given by

$$R_{\ddot{U}}(\tau) = \frac{W_{\ddot{U}}}{2}\delta(\tau), \tag{6.4}$$

where $W_{\ddot{U}}(g^2/Hz)$ is the white noise PSD in the cyclic frequency domain. The correlation matrix of Y_1, Y_2 is given by

$$[R_{Y_1Y_2}(0)] = [R_{Y_1Y_2}] = \begin{pmatrix} < Y_1Y_1 > & < Y_1Y_2 > \\ < Y_2Y_1 > & < Y_2Y_2 > \end{pmatrix} = \begin{pmatrix} < Y_1Y_1 > & 0 \\ 0 & < Y_2Y_2 > \end{pmatrix}, \tag{6.5}$$

where $< Y_i Y_j >$ is the expected value. The process is assumed to be stationary; thus, $\frac{d}{dt} < Y_1 Y_1 >=< Y_2 Y_1 > + < Y_1 Y_2 >= 0$. This means that $< Y_2 Y_1 >= - < Y_1 Y_2 >= 0$.

The Lyapunov equation becomes

$$
\begin{aligned}
&[A][R_{Y_1 Y_2}] + [R_{Y_1 Y_2}][A]^T = -\{B\}\frac{W_{\ddot{U}}}{2}\{B\}^T \\
&\begin{bmatrix} 0 & 1 \\ -\omega_n^2 & -2\zeta\omega_n \end{bmatrix} \begin{bmatrix} < Y_1 Y_1 > & 0 \\ 0 & < Y_2 Y_2 > \end{bmatrix} + \begin{bmatrix} < Y_1 Y_1 > & 0 \\ 0 & < Y_2 Y_2 > \end{bmatrix} \begin{bmatrix} 0 & -\omega_n^2 \\ 1 & -2\zeta\omega_n \end{bmatrix} \\
&= \begin{bmatrix} 0 & 0 \\ 0 & -\frac{W_{\ddot{U}}}{2} \end{bmatrix}.
\end{aligned}
\tag{6.6}
$$

The solution of Eq. (6.6) is

$$
\begin{aligned}
< Y_1 Y_1 > &= E\{Z^2\} = \frac{W_{\ddot{U}}}{8\zeta\omega_n^3} = \frac{W_{\ddot{U}}}{64\pi^3\zeta f_n^3}, \\
< Y_2 Y_2 > &= E\{\dot{Z}^2\} = \frac{W_{\ddot{U}}}{8\zeta\omega_n} = \frac{W_{\ddot{U}}}{16\pi\zeta f_n}.
\end{aligned}
\tag{6.7}
$$

The expected value of the spring force and the damper force $\{F_o\}$, $E[\{F_o\}\{F_o\}^T]$, can be obtained from

$$
\begin{aligned}
E[\{F_o\}\{F_o\}^T] = [C][R_{Y_1 Y_2}][C]^T &= \begin{pmatrix} E[F_k^2] & 0 \\ 0 & E[F_c^2] \end{pmatrix} = \begin{pmatrix} < F_k F_k > & 0 \\ 0 & < F_c F_c > \end{pmatrix} \\
&= \begin{pmatrix} \omega_n^4 m^2 < Y_1 Y_1 > & 0 \\ 0 & 4\zeta^2\omega_n^2 m^2 < Y_2 Y_2 > \end{pmatrix} \\
&= \frac{\pi}{2} f_n Q m^2 W_{\ddot{U}} \begin{pmatrix} 1 & 0 \\ 0 & \frac{1}{Q^2} \end{pmatrix},
\end{aligned}
\tag{6.8}
$$

where $Q = 1/2\zeta$.

The r.m.s. values of the spring force and the damper force are, respectively,

$$
\begin{aligned}
F_{k,rms} &= m\sqrt{\frac{\pi}{2} f_n Q W_{\ddot{U}}}, \\
F_{c,rms} &= \frac{m}{Q}\sqrt{\frac{\pi}{2} f_n Q W_{\ddot{U}}}.
\end{aligned}
\tag{6.9}
$$

The r.m.s. damper force is very small compared to the r.m.s. spring force.

Example

Calculate the r.m.s. reaction force of a SDOF system enforced by a random acceleration given in Table 6.1.

Table 6.1 Random vibration specification

Frequency (Hz)	PSD random acceleration (g^2/Hz)
20–80	3 dB/oct.
80–500	0.04
500–2000	−3 dB/oct.
G_{rms}	8.8 g

The natural frequency of the SDOF element is $f_o = 100\,\text{Hz}$. The mass $m = 50\,\text{kg}$. The damping ratio is $\zeta = 0.05$. The structural reaction force is

$$F_{k,rms} = m\sqrt{\frac{\pi}{2} f_n Q W_{\ddot{U}}} = 3888\ \text{N}.$$

6.3 Pressure-Loaded SDOF System

The SDOF damped equation of motion representing the pressure-loaded SDOF and loaded by the force $F = Ap(t)$, is to be expressed in terms of the displacement $X(t)$

$$\ddot{X}(t) + 2\zeta\omega_n \dot{X}(t) + \omega_n^2 Y(t) = \frac{F(t)}{m}. \tag{6.10}$$

Equation (6.10) will be transformed into state space variables $Y_1 = X(t)$ and $Y_2 = \dot{X}(t)$:

$$\begin{aligned}\begin{pmatrix}\dot{Y}_1\\ \dot{Y}_2\end{pmatrix} &= \begin{pmatrix}0 & 1\\ -\omega_n^2 & -2\zeta\omega_n\end{pmatrix}\begin{pmatrix}Y_1\\ Y_2\end{pmatrix} + \begin{pmatrix}0\\ \frac{1}{m}\end{pmatrix}F,\\ \{\dot{y}\} &= [A]\{Y\} + \{B\}F.\end{aligned} \tag{6.11}$$

The spring force F_k and the force in the damper F_c are considered to be the output variables

$$\{F_o\} = \begin{pmatrix}F_k\\ F_c\end{pmatrix} = \begin{pmatrix}k & 0\\ 0 & c\end{pmatrix}\begin{pmatrix}Y_1\\ Y_2\end{pmatrix} = \begin{pmatrix}\omega_n^2 m & 0\\ 0 & 2\zeta\omega_n m\end{pmatrix}\begin{pmatrix}Y_1\\ Y_2\end{pmatrix} = [C]\{y\}. \tag{6.12}$$

The auto correlation function of the applied force F is

$$R_{FF}(\tau) = \frac{W_{FF}}{2}\delta(\tau), \tag{6.13}$$

where $W_{FF}(N^2/Hz)$ is the white noise PSD in the cyclic frequency domain. The correlation matrix of Y_1, Y_2 is given by

$$[R_{Y_1Y_2}(0)] = [R_{Y_1Y_2}] = \begin{pmatrix} <Y_1Y_1> & <Y_1Y_2> \\ <Y_2Y_1> & <Y_2Y_2> \end{pmatrix} = \begin{pmatrix} <Y_1Y_1> & 0 \\ 0 & <Y_2Y_2> \end{pmatrix}, \tag{6.14}$$

where $<Y_iY_j>$ are the expected values. The process is assumed to be stationary, thus $\frac{d}{dt} <Y_1Y_1> = <Y_2Y_1> + <Y_1Y_2> = 0$. This means that $<Y_2Y_1> = - <Y_1Y_2> = 0$.

The Lyapunov equation becomes

$$\begin{aligned} &[A][R_{Y_1Y_2}] + [R_{Y_1Y_2}][A]^T = -\{B\}\frac{W_{FF}}{2}\{B\}^T, \\ &\begin{bmatrix} 0 & 1 \\ -\omega_n^2 & -2\zeta\omega_n \end{bmatrix}\begin{bmatrix} <Y_1Y_1> & 0 \\ 0 & <Y_2Y_2> \end{bmatrix} + \begin{bmatrix} <Y_1Y_1> & 0 \\ 0 & <Y_2Y_2> \end{bmatrix}\begin{bmatrix} 0 & -\omega_n^2 \\ 1 & -2\zeta\omega_n \end{bmatrix} \\ &= \begin{bmatrix} 0 & 0 \\ 0 & -\frac{W_{FF}}{2m^2} \end{bmatrix}. \end{aligned} \tag{6.15}$$

The solution of Eq. (6.15) is

$$\begin{aligned} <Y_1Y_1> = E\{X^2\} &= \frac{W_{FF}}{8\zeta\omega_n^3 m^2} = \frac{W_{FF}}{64\pi^3\zeta f_n^3 m^2} = \frac{A^2 W_{pp}}{64\pi^3\zeta f_n^3 m^2}, \\ <Y_2Y_2> = E\{\dot{X}^2\} &= \frac{W_{FF}}{8\zeta\omega_n m^2} = \frac{W_{FF}}{16\pi\zeta f_n m^2} = \frac{A^2 W_P}{16\pi\zeta f_n m^2}. \end{aligned} \tag{6.16}$$

The expected value of the spring force and the damper force $\{F_o\}$, $E[\{F_o\}\{F_o\}^T]$, can be obtained from

$$\begin{aligned} E[\{F_o\}\{F_o\}^T] = [C][R_{Y_1Y_2}][C]^T &= \begin{pmatrix} E[F_k^2] & 0 \\ 0 & E[F_c^2] \end{pmatrix} = \begin{pmatrix} <F_kF_k> & 0 \\ 0 & <F_cF_c> \end{pmatrix} \\ &= \begin{pmatrix} \omega_n^4 m^2 <Y_1Y_1> & 0 \\ 0 & 4\zeta^2\omega_n^2 m^2 <Y_2Y_2> \end{pmatrix} \\ &= \frac{\pi}{2} f_n Q A^2 W_{pp} \begin{pmatrix} 1 & 0 \\ 0 & \frac{1}{Q^2} \end{pmatrix}, \end{aligned} \tag{6.17}$$

where $Q = 1/2\zeta$.

The rms values of the spring force and the damper force are respectively

$$\begin{aligned} F_{k,rms} &= A\sqrt{\frac{\pi}{2} f_n Q W_{pp}}, \\ F_{c,rms} &= \frac{A}{Q}\sqrt{\frac{\pi}{2} f_n Q W_{pp}}. \end{aligned} \tag{6.18}$$

The r.m.s. damper force is very small compared to the r.m.s. spring force.

Table 6.2 Acoustic qualification test levels DNEPR space launch system

Center frequency f_c (Hz)	SPL (dB) $0dB = 2 \times 10^{-5}$ (Pa)
31.5	125
63	132
125	135
250	134
500	132
1000	132
2000	129
4000	126
8000	121
OASPL	140

Example

SDOF system is exposed to an acoustic pressure field specified in the user's guide of the DNEPR launch vehicle and given in Table 6.2. The exposed area of the SDOF system is $A = 2$ m^2, the qualification factor is $Q = 10$, and the natural frequency is $f_o = 120$ Hz. The bandwidth in the one octave band $\Delta f = \frac{1}{2}\sqrt{2} f_c$. Thus, 120 Hz is contained in the 125 Hz octave band. PSD of the pressure at the center frequency $f_c = 125$ Hz is

$$W_{pp}(125) = \frac{p^2}{\Delta f} = \frac{p_{ref}^2}{\Delta f} 10^{\frac{SPL(125)}{10}} = 143.11 \text{ Pa}^2/\text{Hz}.$$

The r.m.s. spring force can now be calculated

$$F_{k,rms} = A\sqrt{\frac{\pi}{2} f_n Q W_{pp}} = 1038.76 \text{ N}.$$

6.4 Synthesis

The internal spring forces F_s in both SDOF systems will be equated to obtain the 'break even' area/mass ratio (A/m). The area/mass ratio can be calculated using Eqs. (6.9) and (6.18)

$$\frac{A}{m} = \sqrt{\frac{W_{\ddot{U}}}{W_P}} \approx \sqrt{\frac{W_{\ddot{U}}(f_n)}{W_P(f_n)}}. \tag{6.19}$$

Table 6.3 Random vibration test levels

Frequency (Hz)	Levels $W_{\ddot{U}}$ (g^2/Hz)
20–100	+3 dB/octave
100–300	$0.12 \times (m + 20)/(m + 1)$
300–2000	−5 dB/octave

Table 6.4 Acoustic qualification test levels

Center frequency (Hz)	SPL (dB) $0dB = 2 \times 10^{-5}$ Pa
31.5	130
63	135.5
125	139
250	143
500	138
1000	132
2000	128
4000	124
8000	120
OASPL	147

To calculate a possible 'area over mass ratio' a the following SDOF system will be considered:

- The discrete mass is $m = 50$ kg.
- The natural frequency $f_n = 125$ Hz.

In the ECSS standard 'Testing,' [3] qualification levels for the random enforced vibration and acoustic SPLs are recommended. The random vibration specification is shown in Table 6.3 and the SPL in Table 6.4.

PSD of the pressure field at 125 Hz is

$$W_P = \frac{\left(2.0 \times 10^{-5}\right)^2 10^{\frac{SPL(f)}{10}}}{\Delta f} = \frac{\left(2.0 \times 10^{-5}\right)^2 10^{\frac{139}{10}}}{0.7071 \times 125} = 332.88 \, \frac{Pa^2}{Hz}.$$

PSD of the random enforced acceleration is

$$W_{\ddot{U}} = 0.12 \frac{m + 20}{m + 1} = 0.12 \frac{50 + 20}{50 + 1} = 0.16 \frac{g^2}{Hz} = 15.55 \frac{\left(\frac{m}{s^2}\right)^2}{Hz}.$$

The area/mass ratio becomes

$$\frac{A}{m} = \sqrt{\frac{W_{\ddot{U}}(f_n)}{W_P(f_n)}} = \sqrt{\frac{15.55}{332.88}} = 0.216\,\frac{m^2}{kg}.$$

In [1, 2], a typical 'break even' of $A/m = 0.215\,\frac{m^2}{kg}$ is given.

The area/mass ratio of a test object is important in the selection of a vibration test method. Acoustic testing is more effective for high area/mass components, while random vibration is more effective for low area/mass components.

6.5 Chapter Summary

In this chapter, a key factor, the area/mass ratio of an instrument, component, etc., has been derived, which is very helpful to make a choice to perform either a random vibration on a vibration table or an acoustic test in a acoustic chamber.

Problems

6.1 A scientific instrument has a measured weight of 60 kg. The surplus of heat is radiated into outer space with the aid of an external radiator with total area of 0.75 m. The minimum required natural frequency of the instrument fixed at the spacecraft is $f_n = 100\,\text{Hz}$. The random vibration specification is provided in Table 6.5.

- Calculate the 3σ design quasi-static acceleration with the aid of Miles' equation ($Q = 10$).
- Evaluate if a random vibration test or acoustic test is needed to verify design specifications.
- The acoustic loads of the DNEPR launch vehicle are specified in Table 6.6. Verify if the key factor $A/m = 0.251$ is still applicable for $f_n = 100, 200, 300\,\text{Hz}$.

Table 6.5 Spectral density of vibro-accelerations, DNEPR (logarithmic distribution)

Frequency band (Hz)	PSD (ASD) g^2/Hz
20–80	0.007
80–160	0.007–0.022
160–320	0.022–0.035
320–640	0.035
640–1280	0.035–0.017
1280–2000	0.017–0.005
G_{rms}	6.5

Table 6.6 DNEPR acoustic loads

One octave band (Hz)	SPL (dB), $p_{ref} = 2 \times 10^{-5}$ Pa
31.5	125
63	132
125	135
250	134
500	132
1000	129
2000	126
4000	121
8000	115
OASPL	140
Duration	35 s

References

1. Forgrave JC, Man KF, Newell JM (1998) Acoustic and random vibration test tailoring for low-cost missions. In: Proceedings institute of environmental science, Phoenix, AZ
2. Forgrave JC, Man KF, Newell JM (1998) Spacecraft acoustic and random vibration test optimization. Sound Vib 33(3):28–31
3. European Cooperation of Space Engineering, Noordwijk, the Netherlands. Space Engineering Testing, ECSS-E-ST-10-03C, 3rd edn, June 1st 2012

Chapter 7
Preliminary Predictions of Loads Induced by Acoustic Environment

Abstract A simple manner to estimate structural responses for simple structures exposed to acoustic loads is discussed. Peak pressures are obtained using Miles' equation. Shape factors are introduced to include the effect of boundary conditions. Spann's prediction method of the vibrational environment of components mounted on plates or panels is discussed too. Examples and problems with answers are provided.

Keywords Simple response analysis · Acoustic load · Shape factors · Spann's method

7.1 Introduction

In this chapter, a very simple method to predict the r.m.s. acceleration of a plate-like structure exposed to random acoustic loads. The prediction of the responses is based on Miles' equation. The analysis procedure is done step by step. This procedure is introduced by ESA[1]/ESTEC,[2] department structures [1] to calculate the equivalent static load environment of radiant panels.

7.2 Analysis Procedure

The sound-pressure-loaded SDOF system is illustrated in Fig. 7.1. From Eq. (2.35), the following expression for $\ddot{x}_{rms}$ can be derived, which is

$$\ddot{x}_{rms} = (2\pi f_n)^2 x_{rms} = \sqrt{\frac{\pi}{2} f_n Q \left(\frac{A_p^2 W_p(f_n)}{m^2}\right)}, \tag{7.1}$$

[1] European Space Agency.

[2] European Space Research and Technology Center.

J. Wijker, *Miles' Equation in Random Vibrations*, Solid Mechanics and Its Applications 248, https://doi.org/10.1007/978-3-319-73114-8_7

Fig. 7.1 SDOF element loaded by random plane wave sound pressures

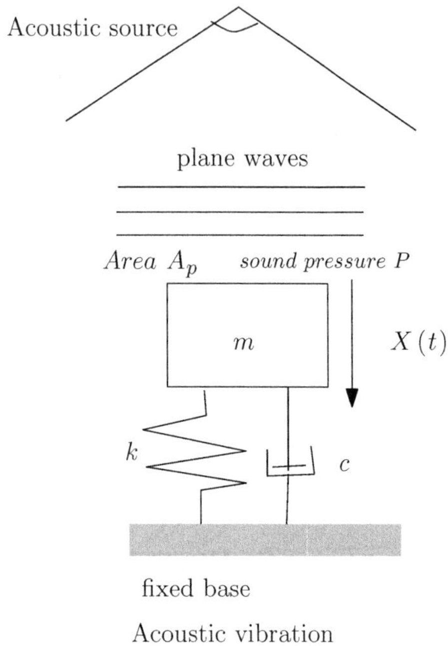

where A_p is the area, m is the mass of the SDOF element, Q is the amplification factor, f_n is the natural frequency of the SDOF element, and W_p is the PSD of the pressure field P. The PSD of the acceleration W_A (g^2/Hz) of the mass m is given by [2, 3]

$$W_A(f_n) = W_p(f_n)\left(\frac{A_p}{mg}\right)^2. \tag{7.2}$$

For a flexible structure, Eq. 7.1 shall be multiplied by a shape factor K_s, but $K_s = 1$ for an unconstrained structure.

The PSD of the random sound pressure $W_p(f_n)$ (Pa^2/Hz) can be calculated from the given $SPL(f_n)$ (dB) at the natural frequency f_n (Hz). The modal damping ratio is ζ, and the amplification factor is $Q = 1/2\zeta$. The PSD function of the pressure P can be calculated by the following expression

$$W_p(f_n) = \frac{p_{ref}^2 10^{\frac{SPL(f_{cent})}{10}}}{\Delta f}, \tag{7.3}$$

where $p_{ref} = 2 \times 10^{-5}$ Pa and Δf the bandwidth $f_{up} - f_{low}$, with an upper frequency f_{up} (Hz) and a lower frequency f_{low} (Hz). The natural frequency f_n is $f_{low} \leq f_n < f_{up}$. The center frequency is given by $f_{cent} = \sqrt{f_{up} f_{low}}$. The ratio $f_{up}/f_{low} = 2^x$, where x stands for one octave or one-third octave band. When f_{cent} (Hz) is the center frequency, the following relations for upper frequency, lower frequency, and

Table 7.1 Frequency bands

x octave	f_{up} (Hz)	$f_{low}(Hz)$	Δf (Hz)
1	$f_{cent}\sqrt{2}$	$f_{cent}/\sqrt{2}$	$\frac{1}{2}\sqrt{2}f_{cent}$
$\frac{1}{3}$	$f_{cent}2^{1/6}$	$f_{cent}/2^{1/6}$	$0.2316f_{cent}$

Table 7.2 Example radiant cooler

Case			Acceptance	Qualification
Acoustic level	SPL	dB	139	142
Frequency	f_n	Hz	216	216
Frequency band	Δf	Hz	176.8	176.8
Pressure density	W_p	Pa^2/Hz	179.7	358.6
Area	A_p	m^2	0.5	0.5
Mass panel	m	kg	3	3
Shape factor	K_s		1	1
Acceleration density	W_A	$g^2/$ Hz	0.052	0.104
r.m.s. acceleration	$\sqrt{\frac{\pi}{2} f_n Q W_A}$	g	13.27	18.74
3σ Acceleration	$3\sqrt{\frac{\pi}{2} f_n Q W_A}$	g	39.80	56.22
Pressure	$\sqrt{\frac{\pi}{2} f_n Q W_p}$	Pa	780.9	1103.1
3σ Pressure	$3\sqrt{\frac{\pi}{2} f_n Q W_p}$	Pa	2342.7	3309.2

bandwidth can be derived and presented in Table 7.1. The results of the calculations of the responses of the radiant cooler exposed to an acoustic environment, both acceptance and qualification levels, are presented in Table 7.2. The 3σ pressure can be applied as a static pressure to investigate strength requirements.

7.3 Evaluation of Shape Factor

For an unconstrained structure, the shape factor $K_s = 1$, but for a flexible structure, the shape factor $K_s \leq 1$. The shape factor will be evaluated for a simply supported rectangular plate with bending stiffness $D = Et^3/(12(1-\nu^2))$ (Nm), length a (m), and width b (m). Young's modulus is E (Pa), Poisson's ratio is ν, and the thickness of plate is t (m). The mass per unit of area of the plate is m_p (kg/m^2).

The equation of motion of the rectangular elastic plate is given by

$$D\left(\frac{\partial^4 W}{\partial x^4}+\frac{\partial^2 W}{\partial x^2}\frac{\partial^2 W}{\partial y^2}+\frac{\partial^4 W}{\partial y^4}\right)+m_p\frac{\partial^2 W}{\partial t^2}=P(t), \tag{7.4}$$

where $W(x, y, t)$ is the random deflection and random P the random plane wave pressure field with a PSD W_p. The solution of W is based on the assumed mode $\phi(x, y)$ multiplied by the generalized coordinate $\eta(t)$. The solution of the deflection $W(x, y, t)$ can be written as

$$W(x, y, t)=\eta(t)\phi(x, y)=\eta(t)\sin\left(\frac{\pi x}{a}\right)\sin\left(\frac{\pi y}{b}\right). \tag{7.5}$$

The corresponding natural frequency f_n (Hz) is given by

$$f_n=\frac{\pi}{2}\sqrt{\frac{D}{m_p}}\left[\left(\frac{1}{a}\right)^2+\left(\frac{1}{b}\right)^2\right]. \tag{7.6}$$

The equation of motion for the generalized coordinate $\eta(t)$, after introducing ad hoc the modal damping ratio ζ, and can now be written

$$\ddot{\eta}+2\zeta\omega_n\dot{\eta}+\omega_n^2\eta=\frac{16}{\pi^2}\frac{A_p P}{M_p}, \tag{7.7}$$

where $A_p = ab$ (m^2) is the total area and $M_p = m_p ab$ (kg) is the total mass of the plate. The r.m.s. value of the acceleration of the generalized coordinate $\ddot{\eta}_{rms}$ is Miles' equation of pressure-loaded SDOF elements

$$\ddot{\eta}_{rms}=\frac{16}{\pi^2}\sqrt{\frac{\pi}{2}f_n Q W_p\left(\frac{A_p}{M_p}\right)^2}. \tag{7.8}$$

The r.m.s. value of the acceleration of the deflection of the plate can be written as

$$\ddot{w}(x, y)_{rms}=\ddot{\eta}_{rms}\phi(x, y)=\frac{16}{\pi^2}\sqrt{\frac{\pi}{2}f_n Q W_p\left(\frac{A_p}{M_p}\right)^2}\left[\sin\left(\frac{\pi x}{a}\right)\sin\left(\frac{\pi y}{b}\right)\right]. \tag{7.9}$$

The average r.m.s. deflection over the area of the plate can be written

$$<\ddot{w}_{rms}>=\frac{1}{A_p}\int_{A_p}\ddot{w}_{rms}dxdy=\frac{64}{\pi^4}\ddot{\eta}_{rms}=0.6570\ddot{\eta}_{rms}. \tag{7.10}$$

Thus, the shape factor of a simply supported plate loaded by a plane wave random pressure field is $K_s = 0.6570$.

For a number of structures, the shape factor K_s is calculated and presented in Table 7.3. For a simply supported circular plate, the shape factor is dependent on Poisson's ratio ν.

Table 7.3 Shape factor for several types of flat structures

Structure type	K_s
Fixed-free beam	0.62
Pinned-pinned	0.62
Simply–simply supported beam	0.81
Simply supported rectangular plate	0.66
Clamped rectangular plate	0.44
Simply supported circular plate	$(21\nu^2+252\nu+756)/(40\nu^2+400\nu+1060)$ $\mu_{K_s}=0.71, 0.0\leq\nu\leq 0.5.$
Clamped circular plate	0.53
Clamped elliptical plate	0.16

7.4 Spann's Component Predictor

Before the Spann's component predictor method will be discussed, the harmonic response of a circular simply supported plate/panel exposed to a one-sided constant over the area acoustic pressure $P(\omega)$ will be solved in the center ($r=0$) of the plate. The equation of motion in the frequency domain of the structural (hysteric) damped plate is given by [4]

$$D(1+jg)\nabla_1^2(\nabla_1^2 W(r,\omega)) - m\omega^2 W(r,\omega) = P(\omega), \tag{7.11}$$

where $j=\sqrt{-1}$, g is the structural damping coefficient, $D=EI$ is the bending stiffness, the nabla operator is given by $\nabla_1^2=\frac{1}{r}\frac{d}{dr}(r\frac{d}{dr})$, W is the deflection, and m is the mass per unit of area. The solution is assumed to be

$$W(r,\omega)=\eta(\omega)\Phi(r), \tag{7.12}$$

where $\eta(\omega)$ is the generalized coordinate and $\Phi(r)$ is the assumed mode [5]

$$\Phi(r)=(4+\nu)a^3-3(2+\nu)ar^2+2(1+\nu)r^2, \tag{7.13}$$

with Poisson's ratio $\nu=0.33$ and $2a$ is the diameter of the circular plate. The undamped natural frequency ω_n can be calculated with

$$\omega_n^2=\frac{D\int_0^a 2\pi r\Phi(r)\nabla_1^2(\nabla_1^2\Phi(r))dr}{m\int_0^a 2\pi r\Phi^2(r)dr}=24.8867\frac{D}{a^4 m}. \tag{7.14}$$

The modal participation factor Γ can be obtained with

$$\Gamma=\frac{\int_0^a 2\pi r\Phi(r)dr}{m\int_0^a 2\pi r\Phi^2(r)dr}=\frac{0.3704}{a^3 m}. \tag{7.15}$$

The damped equation of motion expressed in $\eta(\omega)$ can now be written as

$$\ddot{\eta}(\omega) + \omega_n^2(1 + jg)\eta(\omega) = \Gamma P(\omega). \tag{7.16}$$

The acceleration $\ddot{\eta}(\omega)$ can be calculated using the following expression

$$\ddot{\eta}(\omega) = \frac{-\omega^2}{(\omega_n^2 - \omega^2) + jg\omega_n^2}\Gamma P(\omega). \tag{7.17}$$

The absolute value of the physical acceleration $\ddot{W}$ at the center of the plate ($r = 0$) and at the natural frequency ω_n can be derived from Eq. (7.17)

$$|\ddot{W}(0, \omega_n)| = a(\omega_n) = \frac{1}{g}\Phi(0)\Gamma P(\omega_n) = Q\Phi(0)\Gamma P(\omega_n) = 1.6037\frac{QP(\omega_n)}{m}$$
$$1.6037\frac{QAP(\omega_n)}{M} = \bar{\beta}Q\frac{AP(\omega_n)}{M}, \tag{7.18}$$

where M is the total mass and A is the area of the plate. In case of a simply supported rectangular plate, the factor $\bar{\beta} = 1.6$. The reader is challenged to repeat the response analysis for a simply supported rectangular plate. Equation (7.18) is the basis for an expression in PSD of the acceleration W_a at the center of the plate and PSD of the applied acoustic pressure W_p

$$W_a(\omega_n) = \bar{\beta}^2 Q^2 \left(\frac{A}{M}\right)^2 W_p(\omega_n). \tag{7.19}$$

Equation (7.19) is the starting point of Spann's semiempirical component environment prediction method.

Spann and Patt [6–8] developed a semiempirical method for estimating the response of plates and sandwich panels to the acoustic environment. This response may be used to specify the random environment of the components mounted on the panel. With Spann's method, the acceleration spectral density $W_a(f)$ ((m/s^2)^2/Hz or g^2/Hz) can be calculated using

$$W_a(f) = \beta^2 Q^2 \left(\frac{A}{M}\right)^2 W_p(f), \tag{7.20}$$

where $W_p(f)$ is the PSD of the acoustic pressures (Pa^2/Hz) (so-called SDOF response curve), β is an experimental factor, Q is the amplification factor, A is the area (m^2) of the plate or panel, and M is the total mass (kg) of the panel and mounted equipment as well. This is illustrated in Fig. 7.2. Based on experimental data $\beta = 2.5$ and the amplification factor $Q = 4.5$ are recommended. Later on in 1990, Spann recommended a quality factor $Q = 5$ to ensure that 95% probability level is maintained.

To establish the component random enforced acceleration PSD specification, the SDOF response curve $W_a(f)$ can be adjusted as follows [6]:

- Identify the frequency f_p (Hz) at which the PSD $W_a(f_p)$ has a maximum value.
- Draw a flatline at the $W_a(f_p)$ level between $0.5 f_p$ and $4.0 f_p$.
- Below $0.5 f_p$ roll down the spectrum at 6 dB/oct to 20 Hz.
- Above $4.0 f_p$ roll down the spectrum at 6 dB/oct to 2000 Hz.

Example

A panel with a total area $A = 0.5\,\text{m}^2$ and a total mass $M = 5$ kg is excited by an Ariane 5 specified flight limit (FL) specification, and all other calculations ($\beta = 2.5$, $Q = 5$) are presented in Table 7.4.

The SDOF response curve and the modified response curve (specification) are shown in Fig. 7.3.

7.5 Chapter Summary

In this chapter, a simple manner to estimate structural responses for simple structures when exposed to acoustic loads is discussed. Shape factors will introduce the effect of boundary conditions.

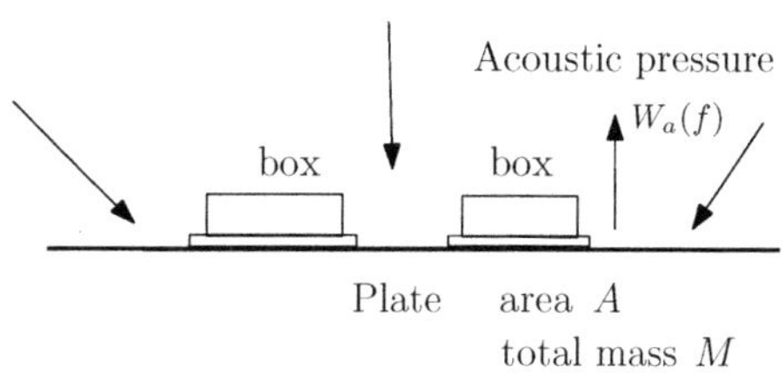

Fig. 7.2 Spann's component predictor

Table 7.4 Spann predictor results

Octave band f_c (Hz)	SPL (dB) $p_{ref} = 2 \times 10^{-5}$ Pa	$\Delta f = 0.7071 f_c$ (Hz)	$p^2_{rms} = p^2_{ref} 10^{\frac{SPL}{10}}$ $\times 10^4$ Pa^2	$W_a(f_c) = p^2_{rms}/\Delta f$ Pa^2/Hz	$W_a(f_c)$ g^2/Hz
31.5	128	22.3	0.2524	113.3090	1.8397
63	131	44.5	0.5036	113.0406	1.8353
125	136	88.4	1.5924	180.1627	2.9251
250	133	176.8	0.7981	45.1476	0.7330
500	129	353.6	0.3177	8.9868	0.1459
1000	123	707.1	0.0798	1.1287	0.0183
2000	116	1414.2	0.0159	0.1126	0.0018
OASPL	139.5 dB		139.4938 dB	G_{rms}	28.5904 g

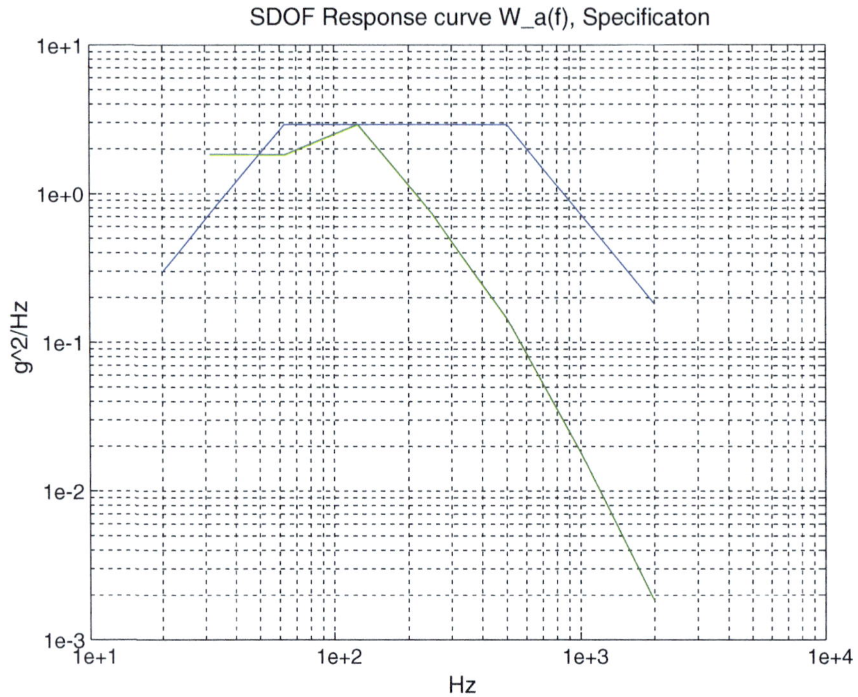

Fig. 7.3 SDOF response curve and specification

Spann's prediction method of the vibrational environment of components mounted on plates or panels is discussed too.

Problems

7.1 Calculate the shape factor K_s for a fixed-free beam with length L and width b, bending stiffness EI, mass per unit of length m_b and loaded by a random plane wave pressure P. The deflection of the beam is $W(x,t)$. The equation of motion is given by

$$EI\frac{\partial^4 W}{\partial x^4} + m_b\frac{\partial^2 W}{\partial t^2} = P(t)b.$$

The assumed vibration mode is [9]

$$\phi(x) = 8((1/4)(x/L)^2 - (1/6)(x/L)^3 + (1/24)(x/L)^4)$$

Answer: $K_s = 0.623$.

7.2 Calculate the shape factor K_s for a clamped rectangular plate with length a and width b, bending stiffness D, mass per unit of area m_p and loaded by a random plane wave pressure P. The deflection of the plate is $W(x, y, t)$. The equation of motion is given by

$$D\left(\frac{\partial^4 W}{\partial x^4}+\frac{\partial^2 W}{\partial x^2}\frac{\partial^2 W}{\partial y^2}+\frac{\partial^4 W}{\partial y^4}\right)+m_p\frac{\partial^2 W}{\partial t^2}=P(t).$$

The assumed vibration mode is [10]

$$\phi(x, y)=(\cos(\pi x/a)-1)(\cos(\pi y/b)-1).$$

Answer: $K_s = 0.444$.

7.3 Calculate the shape factor K_s for a simply supported plate with radius $r = a$, bending stiffness D, mass per unit of area m_p and loaded by a random plane wave pressure P. The deflection of the plate is $W(r, t)$. The equation of motion is given by

$$D\frac{1}{r}\frac{1}{\partial r}\left\{\frac{1}{r}\frac{1}{\partial r}\left[\frac{1}{r}\frac{1}{\partial r}\left(\frac{1}{r}\frac{W(r,t)}{\partial r}\right)\right]\right\}+m_p\frac{\partial^2 W(r,t)}{\partial t^2}=P(t).$$

The assumed vibration mode is [5]

$$\phi(r)=(4+\nu)a^3-3(2+\nu)ar^2+2(1+\nu)r^3.$$

Answer: $K_s = (21\nu^2+252\nu+756)/(40\nu^2+400\nu+1060)$.

7.4 Calculate the shape factor K_s for a clamped plate with radius $r = a$, bending stiffness D, mass per unit of area m_p and loaded by a random plane wave pressure P. The deflection of the plate is $W(r, t)$. The axis symmetric equation of motion is given by

$$D\frac{1}{r}\frac{1}{\partial r}\left\{\frac{1}{r}\frac{1}{\partial r}\left[\frac{1}{r}\frac{1}{\partial r}\left(\frac{1}{r}\frac{W(r,t)}{\partial r}\right)\right]\right\}+m_p\frac{\partial^2 W(r,t)}{\partial t^2}=P(t).$$

The assumed vibration mode is [5]

$$\phi(r)=a^3-3ar^2+2r^3.$$

Answer: $K_s = 0.525$.

7.5 Calculate the shape factor K_s for a clamped elliptical plate with parameters a and width b, bending stiffness D, mass per unit of area m_p and loaded by a random plane wave pressure P. The deflection of the plate is $W(x, y, t)$. The clamped elliptical plate is shown Fig. 7.4. The equation of motion is given by

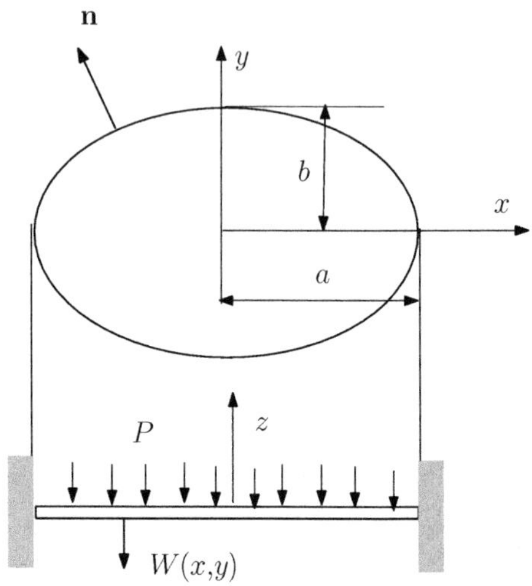

Fig. 7.4 Clamped elliptical plate

$$D\left(\frac{\partial^4 W}{\partial x^4}+\frac{\partial^2 W}{\partial x^2}\frac{\partial^2 W}{\partial y^2}+\frac{\partial^4 W}{\partial y^4}\right)+m_p\frac{\partial^2 W}{\partial t^2}=-P(t).$$

The assumed vibration mode is [11]

$$\phi(x,y)=((x/a)^2+(y/b)^2-1)^2.$$

In addition calculate with aid of the Rayleigh quotient the natural frequency f_n (Hz). Why is along the edge of the ellipse the derivative $dw/dn=0$?
Answer: $K_s=0.159$. $f_n=0.2093\frac{\sqrt{24a^4+16a^2b^2+24b^4}}{a^2b^2}\sqrt{\frac{D}{m_p}}$ (Hz)

7.6 This problem is based on an example discussed in a paper of Spann and Patt [6]. Several components are supported on a sandwich panel with area $A=0.518\,\text{m}^2$. The mass of the panel is $M_p=13\,\text{kg}$. The total mass of the components is $M_c=122\,\text{kg}$. The panel is supported by a truss frame in a S/C which will be exposed to a one-third octave specified sound field given in Table 7.5. Calculate the OASPL.

Perform the following steps:

1. Convert the acoustic environment (Table 7.5) into pressure PSD $W_p(f)$ (Pa^2/Hz).
2. Calculate the ratio A/Mg ($g=9.81$ m/s^2) using the total mass of the assembly ($M=M_p+M_c$).
3. Using the equation $W_a(f)=126.6\left(\frac{A}{Mg}\right)^2 W_p(f)$ (g^2/Hz) to construct the SDOF response curve in the one-third octave band, calculate G_{rms} (g).
4. Establish the component random enforced acceleration PSD specification by adjusting the SDOF response curve $W_a(f)$.
   - Identify the frequency f_p (Hz) at which the PSD $W_a(f_p)$ has a maximum value.

Table 7.5 S/C acoustic test levels

one-third octave band f_c Hz	31.5	40	50	63	80	100	125	160	200	250
SPL (dB) $p_{ref} = 2 \times 10^{-5}$ Pa	120	123	125	126	128	129	130	130	130	130
one-third octave band f_c Hz	315	400	500	630	800	1000	1250	1600	2000	2500
SPL (dB) $p_{ref} = 2 \times 10^{-5}$ Pa	130	130	128	125	123	121	120	118	117	115

- Draw a flatline at the $W_a(f_p)$ level between $0.5f_p$ and $4.0f_p$.
- Below $0.5f_p$ roll down the spectrum at 6 dB/oct to 20 Hz.
- Above $4.0f_p$ roll down the spectrum at 6 dB/oct to 2000 Hz.
- Discuss and calculate G_{rms} (g).

References

1. Labruyere G (1996) Preliminary predictions of the loads induced by the acoustic environment on the door of the radiant cooler of sciamachy. FAX, April 25 1996
2. Chung YT, Foist BL (1995) Prediction of of payload random vibration loads. In: Proceeding IMAX-XIII. Society of experimental mechanics, pp 934–940
3. Leung K, Foist BL (1995) Prediction of acoustically induced random vibration loads for shuttle payloads. In: AIAA, (AIAA-95-1200-CP), pp 362–366
4. Prescott J (1961) Applied elasticity. Dover Publications Inc., New York
5. Temple G, Bickley WG (1956) Rayleigh's principle and applications to engineering. Dover Publications, New York
6. Spann F, Patt D (1984) Component vibration environment predictor. J Environ Sci XXVII(2):19–24
7. Lalanne C (2002) Mechanical vibration and shock, Random vibration, vol III. HPS. ISBN 1-9039-9605-8
8. Schaefer ED (1997) Evaluation the vibro-acoustic response of honeycomb panels. In: Stechner JL (ed) 19th space simulation conference cost effective testing for the 21st century. NASA Goddard Space Flight Center, NASA, 29–31 Oct 1997
9. Ludolph G, Potma AP, Legger RJ (1963) Sterkteleer. Number Tweede deel in Leerboek der mechanica. Wolters, Groningen, achttiende druk edition
10. Leissa AW (1969) Vibration of plates. NASA, Washington
11. Sechler EE (1968) Elasticity in engineering. Dover Publications Inc., New York

Chapter 8
Dynamic Response of Shell Structures to Random Acoustic Excitation, SDOF Approximation

Abstract Three approximate methods to calculate the responses of shell structures when exposed to an acoustic pressure field using a single degree of freedom (SDOF) system approach are discussed in this chapter. The most straightforward method is the approximation applying the second approach, when the acoustic pressure field is proportional to the assumed or vibration mode of the shell structure. To gain more insight examples and problems with answers are provided.

Keywords Response of shell structures · SDOF approximation · Assumed mode · Vibration mode

8.1 Introduction

In this chapter responses of thin-walled shell structures, exposed to an random acoustic pressure field, are approximated [1]. The dynamic behavior of the shell structure is represented by a SDOF element. The acoustic pressure field is separated in spatial and time domain. Three variations of the spatial distribution of the acoustic pressure field are discussed:

1. The acoustic field is constant.
2. The acoustic field corresponds with the fundamental vibration mode of the shell structure.
3. The acoustic field is neither constant or corresponds with the fundamental vibration mode of the shell structure.

A few worked out examples are provided and the chapter ended with a number of problems.

J. Wijker, *Miles' Equation in Random Vibrations*, Solid Mechanics and Its Applications 248, https://doi.org/10.1007/978-3-319-73114-8_8

8.2 SDOF Approximation

The normal deflection $w(\mathbf{x}, t)$ of a linear shell structure (plate, curved panel, cylinder) can be expressed in series of vibration modes $\phi_n(\mathbf{x})$ and associated generalized coordinates $\eta_n(t)$. The vibration modes fulfill the boundary conditions.

$$w(\mathbf{x}, t) = \sum_{n=1}^{M} \eta_n(t)\phi_n(\mathbf{x}), \tag{8.1}$$

where $\mathbf{x} = (x_1, x_2)$ is the two-dimensional location vector. This is illustrated in Fig. 8.1.

The orthogonal relation of the vibration mode ϕ_n is given by

$$\int_{A_0} m(\mathbf{x})\phi_n^2(\mathbf{x})d\mathbf{x} = 1, \tag{8.2}$$

where $m(\mathbf{x})$ the mass per unit of area and A_0 represents the mid-surface of the panel. The random generalized force $F_n(t)$ associated with the vibration mode $\phi_n(\mathbf{x})$ is given by the following expression

$$F_n(t) = \int_{A_0} p(\mathbf{x}, t)\phi_n(\mathbf{x})d\mathbf{x}, \tag{8.3}$$

where $p(\mathbf{x}, t)$ is the random distributed pressure. The partial equation of motion of the linear shell can be transferred into SDOF equations applying Eqs. (8.2, 8.3). The damping is introduced as the modal damping ratio ζ_n. The modal equations can be expressed by the generalized coordinates $\eta_n(t)$.

$$\ddot{\eta}_n(t) + 2\zeta_n\omega_n\dot{\eta}_n(t) + \omega_n^2\eta_n(t) = F_n(t), n = 1, 2, \ldots, N, \tag{8.4}$$

where ω_n is the undamped natural frequency of the linear shell structure.

Example

A circular plate with radius a is clamped along the edge and loaded by a uniform dynamic pressure $p(t)$. The undamped axe-symmetric equation of motion is given by [2]

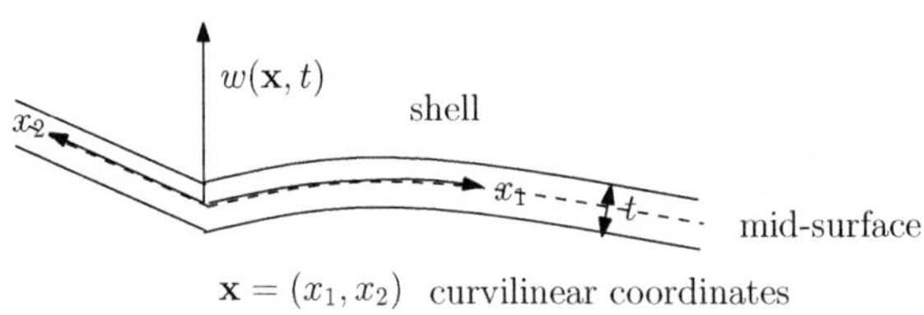

Fig. 8.1 Shell configuration

$$D\frac{1}{r}\frac{d}{dr}\left(r\frac{dw(r,t)}{dr}\right) + m_p\ddot{w}(r,t) = p(t),$$

where $w(r, t)$ is the normal deflection, D is the bending rigidity, r is the running radius, and m_p the constant mass per unit of area. The assumed vibration mode $\phi(r)$ is given by

$$\phi(r) = \frac{\sqrt{35}\,\left(2\,r^3 - 3\,a\,r^2 + a^3\right)}{\sqrt{6}\,\sqrt{\pi}\,a^4\,\sqrt{m_p}},$$

with

$$\int_0^a 2\pi r m_p \phi^2(r) dr = 1.$$

The associated natural frequency ω_n can be calculated with the Rayleigh's quotient [3, 4]

$$\omega_n = \frac{\sqrt{105}\sqrt{\frac{D}{m_p}}}{a^2},$$

and the generalized force F_n is given by

$$F_n = \frac{0.3\,\sqrt{35}\,\sqrt{\pi}\,a\,p(t)}{\sqrt{6}\,\sqrt{m_p}}$$

The deflection $w(r, t)$ is written as

$$w(r,t) = \eta(t)\phi(r).$$

When the damping is introduced in an ad hoc manner the equation of motion expressed in the generalized coordinate $\eta(t)$ making use of Eq. (8.2).

$$\ddot{\eta}(t) + 2\zeta\omega_n\dot{\eta}(t) + \omega_n^2\eta(t) = F_n(t),$$

where ζ is the damping ratio.

The pressure $p(\mathbf{x}, t)$ applied to the shell varies in space and time and can be written as

$$p(\mathbf{x}, t) = \psi(\mathbf{x})p(t), \tag{8.5}$$

where $\psi(\mathbf{x})$ defines the spatial distribution and $p(t)$ the pressure which varies in time. The assumption of a pressure field with separable spatial and temporal characteristics in Eq. (8.5) is only true for certain fields such as reverberant (diffuse) sound fields. Defining $W_p(\mathbf{x}, f)$ as the PSD of the $p(\mathbf{x}, t)$ and $W_o(f)$ as the PSD of pressure $p(t)$, the PSD of $W_p(\mathbf{x}, f)$ becomes

$$W_p(\mathbf{x}, f) = \psi^2(\mathbf{x})W_o(f). \tag{8.6}$$

The space average of $W_p(\mathbf{x}, f)$ is than as follows

$$\langle W_p(\mathbf{x}, f)\rangle_{\mathbf{x}} = \langle \psi^2(\mathbf{x})\rangle_{\mathbf{x}} W_o(f). \tag{8.7}$$

In a special case of a homogeneous pressure field $\langle W_p(\mathbf{x}, f)\rangle_{\mathbf{x}}$ is equal to the PSD measured as a single point. In other cases, it is defined as the space average of the measured PSD, with the measurements taken over a range of points on the panel.

Substituting Eq. (8.5) into Eq. (8.3) an expression for the generalized force $F_n(t)$ as function of the pressure loading, vibration mode, and generalized mass (Eq. (8.2)) is obtained

$$F_n(t) = \frac{\int_{A_o} p(\mathbf{x})\phi_n(\mathbf{x})d\mathbf{x}}{\int_{A_0} m(\mathbf{x})\phi_n^2(\mathbf{x})d\mathbf{x}} = p(t)\int_{A_o} \psi(\mathbf{x})\phi_n(\mathbf{x})d\mathbf{x}, \tag{8.8}$$

where $\int_{A_0} m(\mathbf{x})\phi_n^2(\mathbf{x})d\mathbf{x} = 1$ is the unit generalized mass.

The PSD of the generalized force $F_n(t)$ becomes

$$\begin{aligned} W_{F_n}(f) &= W_o(f)\left[\frac{\int_{A_o} \psi(\mathbf{x})\phi_n(\mathbf{x})d\mathbf{x}}{\int_{A_0} m(\mathbf{x})\phi_n^2(\mathbf{x})d\mathbf{x}}\right]^2 = \frac{\langle W_p(\mathbf{x}, f)\rangle_{\mathbf{x}}}{\langle \psi^2(\mathbf{x})\rangle_{\mathbf{x}}}\left[\frac{\int_{A_o} \psi(\mathbf{x})\phi_n(\mathbf{x})d\mathbf{x}}{\int_{A_0} m(\mathbf{x})\phi_n^2(\mathbf{x})d\mathbf{x}}\right]^2 \\ &= \frac{\langle W_p(\mathbf{x}, f)\rangle_{\mathbf{x}}}{\langle \psi^2(\mathbf{x})\rangle_{\mathbf{x}}}\left[\int_{A_o} \psi(\mathbf{x})\phi_n(\mathbf{x})d\mathbf{x}\right]^2, \end{aligned} \tag{8.9}$$

where $\langle W_p(\mathbf{x}, f)\rangle_{\mathbf{x}}$ is averaged measured PSD of the pressure, $\phi_n(\mathbf{x})$ is the response vibration mode, and $m(\mathbf{x})$ the mass per unit of area, as before. It therefore remains to specify $\psi(\mathbf{x})$ in order to fully define the broadband random excitation spectrum $W_o(f)$. There are several approaches:

Case 1 The spatial distribution function is $\psi(\mathbf{x}) = 1$.
Case 2 The spatial distribution function is $\psi(\mathbf{x}) = \phi_n(\mathbf{x})$.
Case 3 The spatial distribution function is $\psi(\mathbf{x}) \neq \phi_n(\mathbf{x})$.

8.2.1 Case 1, $\psi(x) = 1$

This is the case of uniform pressure loading. Substituting the spatial distribution function $\psi(\mathbf{x}) = 1$ in to Eq. (8.9) will produce

$$W_{F_n}(f) = W_o(f)\left[\frac{\int_{A_o} \phi_n(\mathbf{x})d\mathbf{x}}{\int_{A_0} m(\mathbf{x})\phi_n^2(\mathbf{x})d\mathbf{x}}\right]^2. \tag{8.10}$$

The SDOF system associated with the vibration mode $\phi_n(\mathbf{x})$, the generalized coordinate $\eta_n(t)$ and assuming a unit uniform pressure $p(t) = 1$ can be written as follows

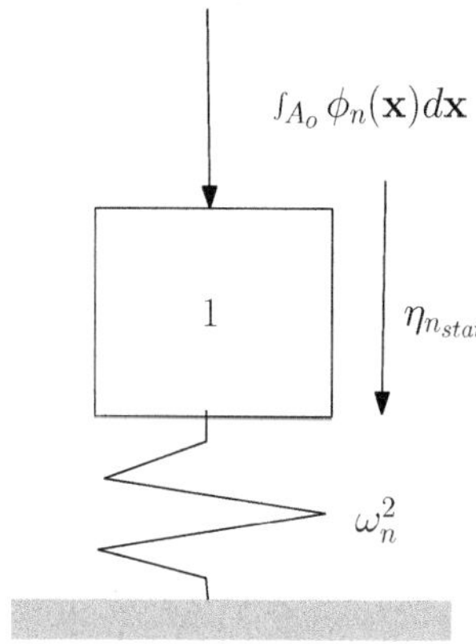

Fig. 8.2 Static displacement SDOF

$$\ddot{\eta}(t) + 2\zeta_n\omega_n\dot{\eta}_n(t) + \omega_n^2\eta_n(t) = \left[\frac{\int_{A_o}\phi_n(\mathbf{x})d\mathbf{x}}{\int_{A_0} m(\mathbf{x})\phi_n^2(\mathbf{x})d\mathbf{x}}\right] = \int_{A_o}\phi_n(\mathbf{x})d\mathbf{x}. \tag{8.11}$$

In the static case $\ddot{\eta}_n(t) = \dot{\eta}_n(t) = 0$ the static displacement of $\eta_{n_{stat}}$ is given by

$$\eta_{n_{stat}} = \frac{1}{\omega_n^2}\int_{A_o}\phi_n(\mathbf{x})d\mathbf{x}, \tag{8.12}$$

where ω_n^2 represents the generalized stiffness. This is illustrated in Fig. 8.2.

Equation (8.10) can now expressed in $\eta_{n_{stat}}$

$$W_{F_n}(f) = W_o(f)\eta_{n_{stat}}^2\omega_n^4. \tag{8.13}$$

The r.m.s. response of the generalized coordinate $\eta_n(t)$ of the SDOF system represented by Eq. (8.4) and loaded by the random load $F_n(t)$ with the PSD $W_{F_n}(f)$ is given by

$$\eta_{n_{rms}} = \sqrt{\frac{W_{F_n}(f_n)}{8\zeta_n(2\pi f_n)^3}} = \eta_{n_{stat}}\sqrt{\frac{\pi}{2}f_n Q_n W_o(f_n)}, \tag{8.14}$$

where the amplification factor $Q_n = 1/2\zeta_n$ and $W_o(f_n)$ the PSD of the random pressure at a natural frequency f_n.

The physical r.m.s. response $w(\mathbf{x})_{rms}$ can be calculated by using Eq. (8.1)

$$w(\mathbf{x})_{rms} = \eta_{n_{rms}}\phi(\mathbf{x}). \tag{8.15}$$

Example

The cantilevered beam with width b is loaded by an uniform random pressure $p(t)$ with a PSD $W_o(f)$. All other parameters and design variables are depicted in Fig. 8.3. m_b is mass of the beam per unit of area.

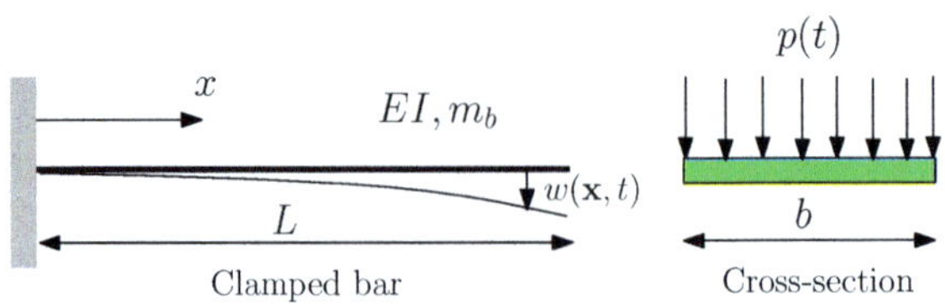

Fig. 8.3 Clamped beam pressure loaded

The assumed vibration mode $\phi_n(x)$ is taken from [4]

$$\phi_n(x) = x^2\left(6\,L^2 - 4\,x\,L + x^2\right),$$

and associated with $A_o = bL$

$$\int_{A_o} \phi_n(x)dx = 1.2\,b\,L^5,$$

and

$$\int_{A_o} m_b\phi_n^2(x)dx = 2.3111\,b\,m_b\,L^9.$$

The estimated natural frequency ω_n^2 is

$$\omega_n^2 = \frac{12.4615\,EI}{m_b b\,L^4},$$

but the theoretical natural frequency [5]

$$\omega_n^2 = \frac{12.3624\,EI}{m_b b\,L^4}.$$

Thus, the approximation is within 1%. When using the estimated natural frequency the static displacement $\eta_{n_{stat}}$ becomes

$$\eta_{n_{stat}} = \frac{1}{\omega_n^2}\left[\frac{\int_{A_o}\phi_n(x)dx}{\int_{A_0} m_b\phi_n^2(x)dx}\right] = \frac{0.04167}{EI}.$$

Example

This example is based on a paper of Dhainaut et al., [6]. Consider the linear vibrations of a simply supported rectangular isotropic plate subjected to a random uniformly distributed pressure $P(t)$ with PSD $W_p(f)$. The transverse displacement $W(x, y, t)$ of a $a \times b \times t$ (length x width x thickness) panel can be approximated as

$$W(x, y, t) = \eta(t)\Phi(x, y) = \eta(t)\sin\left(\frac{\pi x}{a}\right)\sin\left(\frac{\pi y}{b}\right), \tag{8.16}$$

where $\eta(t)$ is the generalized coordinate and $\Phi(x, y)$ the assumed mode. The equation of motion of the simply supported plate is given by

$$D\left[\frac{\partial^4 W}{\partial x^4}+2\frac{\partial^2 W}{\partial x^2}\frac{\partial^2 W}{\partial y^2}+\frac{\partial^4 W}{\partial y^4}\right]+m\frac{\partial^2 W}{\partial t^2}=P(t), \tag{8.17}$$

where D is bending stiffness and m the constant mass per unit area.

Calculate the static generalized displacement η_{stat} using Eq. 8.12. The natural frequency ω_n^2, f_n and $\int_{A_o}\Phi(x, y)dA$ are given by the following expressions:

- $\omega_n^2=\frac{\pi^4\left(b^2+a^2\right)^2 D}{a^4b^4m}$,
- $f_n=\frac{\omega_n}{2\pi}$,
- $\int_{A_o}\Phi(x, y)dA=\frac{4ab}{\pi^2}$

The static displacement η_{stat} can be obtained

$$\eta_{stat}=\frac{1}{\omega_n^2}\frac{\int_{A_o}\Phi(x, y)dA}{m\int_{A_o}\Phi^2(x, y)dA}=\frac{16a^4b^4}{\pi^6\left(b^2+a^2\right)^2 D}. \tag{8.18}$$

The r.m.s. value of the random displacement $W(x, y, t)$, $w(x, y)_{rms}$ can now be calculated with Eqs. (8.14) and (8.15)

$$w(x, y)_{rms}=\eta_{stat}\Phi(x, y)\sqrt{\frac{\pi}{2}f_n Q W_p(f_n)}, \tag{8.19}$$

where Q is the amplification factor. The solution method used for this example is quite similar to the solution method as discussed in Sect. 3.4.

The bending moments M_x, M_y can be calculated using the following expressions

$$M_x=D\left(\frac{\partial^2 W}{\partial x^2}+\nu\frac{\partial^2 W}{\partial y^2}\right),\ M_y=D\left(\frac{\partial^2 W}{\partial y^2}+\nu\frac{\partial^2 W}{\partial x^2}\right). \tag{8.20}$$

8.2.2 Case 2, $\psi(x)=\phi_n(x)$

The pressure field match with the vibration mode $\phi_n(\mathbf{x})$ and this vibration mode is normalized such that

$$\int_{A_0} m(\mathbf{x})\phi_n^2(\mathbf{x})d\mathbf{x}=1.$$

Equation 8.9 can now rewritten as

$$W_{F_n}(f)=\frac{\left\langle W_p(\mathbf{x}, f)\right\rangle_{\mathbf{x}}}{\left\langle\psi^2(\mathbf{x})\right\rangle_{\mathbf{x}}}\left[\frac{\int_{A_o}\psi(\mathbf{x})\phi_n(\mathbf{x})d\mathbf{x}}{\int_{A_0}m(\mathbf{x})\phi_n^2(\mathbf{x})d\mathbf{x}}\right]^2=W_o(f)\left[\int_{A_o}\phi_n^2(\mathbf{x})d\mathbf{x}\right]^2, \tag{8.21}$$

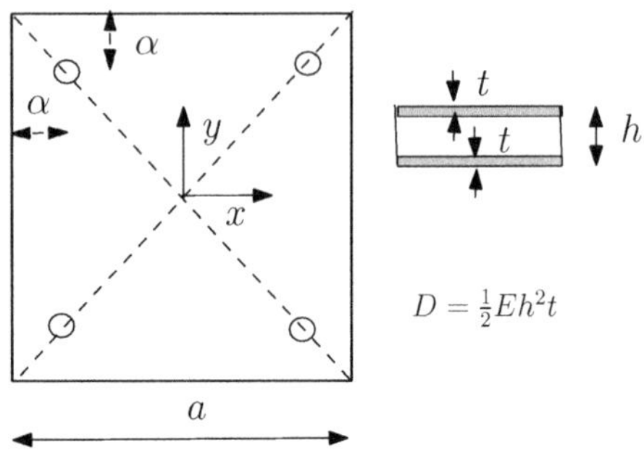

Fig. 8.4 Square solar panel supported symmetrically on four points

where $W_o(f) = mA_o \langle W_p(\mathbf{x}, f)\rangle_{\mathbf{x}}$ and the mass per unit of area $m(\mathbf{x}) = m$ is constant equation 8.21 becomes

$$W_{F_n}(f) = \langle W_p(\mathbf{x}, f)\rangle_{\mathbf{x}} \frac{A_o}{m} = W_o(f)\frac{1}{m^2} = W_o(f)\left(\frac{A_o}{M}\right)^2, \tag{8.22}$$

where M is the total mass of the shell.

The r.m.s. response of the generalized coordinate $\eta_n(t)$ of the SDOF system represented by Eq. (8.4) and loaded by the random load $F_n(t)$ with the PSD $W_{F_n}(f)$ is given by

$$\begin{aligned}\eta_{n_{rms}} &= \sqrt{\frac{W_{F_n}(f_n)}{8\zeta_n(2\pi f_n)^3}} = \frac{1}{(2\pi f_n)^2}\sqrt{\frac{A_o}{m}}\sqrt{\frac{\pi}{2} f_n Q_n \langle W_p(\mathbf{x}, f_n)\rangle_{\mathbf{x}}},\\ &= \frac{1}{(2\pi f_n)^2}\frac{A_o}{M}\sqrt{\frac{\pi}{2} f_n Q_n W_o(f_n)},\end{aligned} \tag{8.23}$$

where the amplification factor $Q_n = 1/2\zeta_n$ and $W_o(f_n)$ the PSD of the random pressure at a natural frequency f_n.

The physical r.m.s. response $w(\mathbf{x})_{rms}$ can be calculated by using Eq. (8.1)

$$w(\mathbf{x})_{rms} = \eta_{n_{rms}}\phi(\mathbf{x}).$$

Example

A square solar panel is symmetrically supported by four points as illustrated in Fig. 8.4.

The solar panel is made of a sandwich construction with Al-alloy face sheets with a thickness $t = 0.5$ mm and the Young's modulus is $E = 70$ GPa. The height of the sandwich panel is h mm. The supports are located α mm from the edges of the panel. The stiffness of the core shall be neglected. Define h such that the lowest natural frequency $f_n \geq 60$ Hz using the following information

- The lowest natural frequency $\omega_n a^2\sqrt{m/D} = 12.58$, $\alpha = 0.1a$, $\nu = 0.33$, [7].

- The mass per unit of area is $m = 2.5\,\text{kg/m}^2$
- The length $a = 1.5\,\text{m}$.

Answer: $h = 25.5\,\text{mm}$.

Calculate the r.m.s. acceleration of the generalized coordinate η_n using Eq. (8.23). The pressure field is proportional to the vibration mode ϕ_n, the modal damping ratio is $\zeta = 0.05$, and the PSD of $W_o(f_n) = 50\,\text{Pa}^2/\text{Hz}$.

$$\ddot{\eta}_{n_{rms}} = \frac{A_o}{M}\sqrt{\frac{\pi}{2} f_n Q_n W_o(f_n)} = 86.83\,\text{m/s}^2.$$

The assumed mode method in combination the Rayleigh's quotient is applied to obtain the lowest natural frequency ω_n. The assumed mode applied is

$$\phi_n(x, y) = B\left(2 - \frac{2^n k^n x^n}{a^n} - \frac{2^n k^n y^n}{a^n}\right), \tag{8.24}$$

where $k = a/(a - 2\alpha)$ and $n = 2$, [7]. The constant B can be obtained using the orthogonality property of vibration modes ($\int_{A_0} m(\mathbf{x})\phi_n^2(\mathbf{x})d\mathbf{x} = 1$) and is given by

$$B = \frac{0.1164}{a\sqrt{m}}.$$

8.2.3 Case 3, $\psi(x) \neq \phi_n(x)$

The spatial distribution function $\psi(\mathbf{x})$ of the random pressure loading is neither uniform nor equal to the vibration mode $\phi_n(\mathbf{x})$. The vibration are such scaled that the generalized mass is equal to unit, thus

$$\int_{A_0} m(\mathbf{x})\phi_n^2(\mathbf{x})d\mathbf{x} = 1.$$

The average mean square of the spatial distribution function can be expressed by

$$\langle\psi^2(\mathbf{x})\rangle = \frac{1}{A_o}\int_{A_o} \psi^2(\mathbf{x})d\mathbf{x}. \tag{8.25}$$

$$W_{F_n}(f) = \frac{\langle W_p(\mathbf{x}, f)\rangle_{\mathbf{x}}}{\langle\psi^2(\mathbf{x})\rangle_{\mathbf{x}}}\left[\frac{\int_{A_o}\psi(\mathbf{x})\phi_n(\mathbf{x})d\mathbf{x}}{\int_{A_0} m(\mathbf{x})\phi_n^2(\mathbf{x})d\mathbf{x}}\right]^2 = \frac{A_o\langle W_p(\mathbf{x}, f)\rangle_{\mathbf{x}}}{\int_{A_o}\psi^2(\mathbf{x})d\mathbf{x}}\left[\frac{\int_{A_o}\psi(\mathbf{x})\phi_n(\mathbf{x})d\mathbf{x}}{\int_{A_0} m(\mathbf{x})\phi_n^2(\mathbf{x})d\mathbf{x}}\right]^2$$
$$= \frac{A_o\langle W_p(\mathbf{x}, f)\rangle_{\mathbf{x}}}{\int_{A_o}\psi^2(\mathbf{x})d\mathbf{x}}\left[\int_{A_o}\psi(\mathbf{x})\phi_n(\mathbf{x})d\mathbf{x}\right]^2, \tag{8.26}$$

The joint acceptance J_n is the degree to which the spatial distribution $\psi(\mathbf{x})$ of pressure loading matches with the vibration mode $\phi_n(\mathbf{x})$ and is defined as

$$J_n = \frac{\int_{A_o} \psi(\mathbf{x})\phi_n(\mathbf{x})d\mathbf{x}}{\sqrt{\int_{A_o} \psi^2(\mathbf{x})d\mathbf{x} \int_{A_o} \phi_n^2(\mathbf{x})d\mathbf{x}}}. \tag{8.27}$$

Equation (8.26) can be finally written

$$W_{F_n}(f) = A_o \langle W_p(\mathbf{x}, f)\rangle_{\mathbf{x}} J_n^2 \int_{A_o} \phi_n^2(\mathbf{x})d\mathbf{x}. \tag{8.28}$$

When the mass per unit of area $m(x) = m$ Eq. (8.28) can be simplified

$$W_{F_n}(f) = \frac{A_o}{m} \langle W_p(\mathbf{x}, f)\rangle_{\mathbf{x}} J_n^2. \tag{8.29}$$

The r.m.s. response of the generalized coordinate $\eta_n(t)$ of the SDOF system represented by Eq. (8.4) and loaded by the random load $F_n(t)$ with the PSD $W_{F_n}(f)$ is given by

$$\eta_{n_{rms}} = \sqrt{\frac{W_{F_n}(f_n)}{8\zeta_n(2\pi f_n)^3}} = \frac{J_n}{(2\pi f_n)^2}\sqrt{\frac{A_o}{m}}\sqrt{\frac{\pi}{2} f_n Q_n \langle W_p(\mathbf{x}, f_n)\rangle_{\mathbf{x}}} \tag{8.30}$$

where the amplification factor $Q_n = 1/2\zeta_n$ and $\langle W_p(\mathbf{x}, f_n)\rangle_{\mathbf{x}}$ the PSD of the averaged random pressure at the natural frequency f_n.

The physical r.m.s. response $w(\mathbf{x})_{rms}$ can be calculated by using Eq. (8.1)

$$w(\mathbf{x})_{rms} = \eta_{n_{rms}}\phi(\mathbf{x}).$$

Example

We proceed with the example illustrated in Fig. 8.3. The spatial distribution along the beam is a unidirectional wave given by

$$\psi(x) = \exp(-jk_a x) \tag{8.31}$$

The joint acceptance J_n^2 can be numerically approximated by the following expression

$$J_n^2 = \frac{|\sum_{k=1}^N \psi(x_k)\phi_n(x_k)|^2}{\sum_{k=1}^N |\psi(x_k)|^2 \sum_{k=1}^N \phi_n^2(x_k)}, \tag{8.32}$$

where $x_1 = 0, \ldots, x_N = L$.

The length of the beam is $L = 1$ m, the natural frequency of the clamped beam $f_n = 50, 250$ Hz and the speed of sound in air $c = 340$ m/s. The vibration mode

Table 8.1 Computation of J_n, $f_n = 50, 250\,\text{Hz}$

N	$f_n = 50\,\text{Hz}$	$f_n = 250\,\text{Hz}$
	J_n	J_n
5	0.7458	0.4528
10	0.7588	0.4734
100	0.7735	0.4999
1000	0.7752	0.5030
10000	0.7753	0.5033

ϕ_n is already provided for the clamped beam. The wavenumber $k_a = \omega_n/c = 0.9240, 4.6200\,\text{m}^{-1}$. The wavelength $\lambda = c/f_n = 6.8000, 1.3600\,\text{m}$.

The numerical calculations are given in Table 8.1.

The shorter the wavelength λ the less the joint acceptance J_n. The convergence rate of the J_n depends more or less of the number of the discrete points x_k or the nodal points in the finite element model (FEM).

8.3 Chapter Summary

Three approximate methods to calculate the responses of shell structures exposed to a acoustic pressure field using a SDOF system has been discussed. The most straightforward method is the approximation applying the CASE#2 approach, when the acoustic pressure field is proportional to the assumed or vibration mode of the shell structure.

Problems

8.1 The scaled vibration mode $\phi(x, y)$ is given by

$$\phi(x, y) = \frac{1}{\sqrt{m_p ab}} \sin\left(\frac{\pi x}{a}\right) \sin\left(\frac{\pi y}{b}\right),$$

where m_p is the constant mass per unit of area of the rectangular panel and the mid-surface is given by $A_0 = ab$. Show that

$$\int_{A_0} m_p \phi^2(x, y)(\mathbf{x}) dx dy = 1.$$

8.2 A circular plate with simply supported edge has a diameter $d = 2a$ and a constant mass m_p per unit of area. The vibration mode $\phi(r)$ is given by [4]

$$\phi(r) = c[(4+\nu)a^3 - 3(2+\nu)ar^2 + 2(1+\nu)r^3]$$

Calculate factor c such that

$$\int_0^a 2\pi r m_p \phi^2(r)dr = 1,$$

and $\nu = 0.33$ and subsequently calculate the generalized force

$$F_n = \int_0^a 2\pi r p \phi(r)dr,$$

where the pressure p is constant over the area of the plate.
Answers: $c = \pm\frac{0.6245585735670467}{a^4\sqrt{m_p}}$, $F_p = \frac{0.5930183656019108\,a\,p}{\sqrt{m_p}}$

8.3 A circular plate with simply supported edge and a constant mass per unit of area m_p has a diameter $d = 2a$. The spatial distribution $\psi(r)$ and the vibration mode $\phi(r)$ are given by

$$\psi(r) = \phi(r) = (4+\nu)a^3 - 3(2+\nu)ar^2 + 2(1+\nu)r^3$$

Calculate $\langle\psi^2(r)\rangle$ with $\nu = 0.33$. Evaluate Eq. (8.8) when the plate is pressure loaded by $p(t)$.

Answers: $\langle\psi^2(r)\rangle = 5.12724a^6$, $F_n(t) = p(t)/m_p$

8.4 Prove that the joint acceptance $J_n = \omega_n^2 \eta_{stat} m/A_o$ when the spatial distribution function $\psi(\mathbf{x}) = 1$.

8.5 Prove that the joint acceptance $J_n = 1$ when the spatial distribution function $\psi(\mathbf{x}) = \phi_n(\mathbf{x})$.

8.6 A sandwich panel is supported at the four corner points and is illustrated in Fig. 8.5.

The lowest natural frequency of the panel shall be $f_n \geq 50\,\text{Hz}$. Design a sandwich panel with Al-alloy face sheets with thickness $t = 0.3\,\text{mm}$, Young's modulus of the face sheets is $E = 7 \times 10^{10}$ Pa, and the mass per unit of area of the panel is $m = 2\,\text{kg/m}^2$. The shear/bending stiffness of the core shall be neglected.

Calculate the joint acceptance J_n taking into account the following information:

- The assumed vibration mode is $\phi_n(x, y) = \sin\left(\frac{\pi x}{a}\right) + \sin\left(\frac{\pi y}{b}\right)$.
- Dimensions of the panel are $b = 1.5a$, $a = 1$ m.
- Speed of sound in air $c = 340$ m/s.
- The wave number $k_a = \omega_n/c$.
- Spatial distribution of sound field $\psi = \exp(-jk_a(x+y))$, $j = \sqrt{-1}$.
- The lowest natural frequency can be calculated using $\omega_n a^2\sqrt{m/D} = 9.21$ ($b = 1.5a$, $\omega_n = 2\pi f_n$), [8].
- Divide a and b in 100, 1000 increments.

Fig. 8.5 Sandwich panel supported at the four corner points

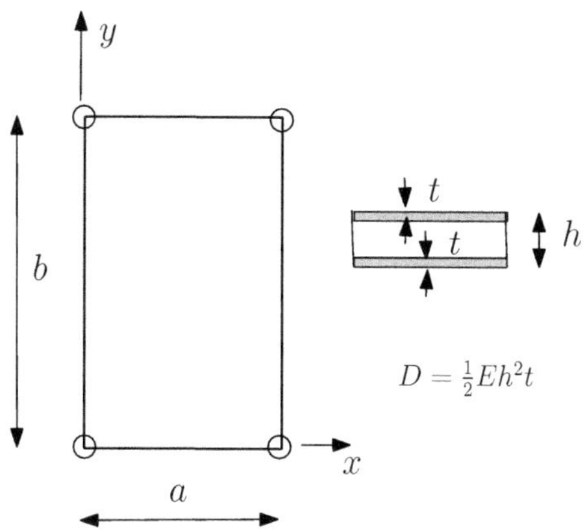

- Apply equation (8.32) in two directions.

Answers: $h = 14.9\,\text{mm}$, $J_n = 0.8600, 0.8631$

8.7 This problem is a continuation of problem 8.6. Calculate the r.m.s. acceleration of the generalized coordinate η_n (m/s^2) applying Eq. (8.30). The modal damping ratio is $\zeta = 0.05$, the averaged PSD of the pressure field $\langle W_p(\mathbf{x}, f_n)\rangle_{\mathbf{x}} = 100\,\text{Pa}^2/\text{Hz}$ and the joint acceptance $J_n = 0.86$.
Answer: $\ddot{\eta}_{rms} = 208.73\,\text{m/s}^2$.

References

1. Cunningham PR, Langley RS, White RG (2003) Dynamic response of doubly curved honeycomb sandwich panels to random acoustic excitation, part 2: theoretical study. J Sound Vib 264:605–637
2. Prescott J (1961) Applied elasticity. Dover Publications Inc., New York
3. Meirovitch L (1980) Computational methods in structural dynamics. Sijthoff and Noordhoff. ISBN 90 286 05800
4. Temple G, Bickley WG (1956) Rayleigh's principle and applications to engineering. Dover Publications, New York
5. Blevins RD (1995) Formulas for natural frequency and mode shape. Krieger Publishing Company. ISBN 0-89464-894-2
6. Dhainaut JM, Cheng G, Mei C (2014) Response of plates under statistically unsynchronized uniform random loads using monte-carlo simulation. J Comput Appl Mech 9(1):3–18
7. Johns DJ, Nagaraj VT (1968) On the fundatmental frequency of square plate symmetrically supported at four points. J Sound Vib 10(3):404–410
8. Leissa AW (1969) Vibration of plates. NASA, Washington

Chapter 9
Equivalence Random and Sinusoidal Vibration

Abstract In general, the sinusoidal vibration (sweep) test is in the frequency range between 5 and 100 Hz and the random vibration test has a frequency range between 20–2000 Hz. Therefore, it is not so straightforward to replace a random vibration test by an equivalent sinusoidal vibration test or vice versa. Damaging and fatigue aspects have to be considered transferring a specific sinusoidal specification into equivalent random vibration specification. Examples and problems are provided.

Keywords Sine vibration test · Random vibration test

9.1 Introduction

The equivalence of the random and sinusoidal response of two SDOF system excited at the base is illustrated in Fig. 9.1. The responses of both SDOF elements are compared, and the r.m.s. (1σ) random response $\sqrt{\frac{\pi}{2} f_n Q_k W_{\ddot{U}}(f_n)}$ is made equal to the r.m.s. of the sinusoidal response $\frac{1}{2} Q_k A_{sine}(f_n)\sqrt{2}$. The equivalent sinusoidal base excitation $A_{sine,1\sigma}(f_n)$ can be obtained by the following expression

$$A_{sine,1\sigma}(f_n) = \sqrt{2}\sqrt{\frac{\pi f_n W_{\ddot{U}}(f_n)}{2Q_k}}, \quad k = 1, 2, \cdots, K. \tag{9.1}$$

When random and sinusoidal input equivalence is investigated the amplification factor $Q = 1$ is taken. In [1], the equivalence of amplitudes between the sinusoidal and random vibration is based on the power spectral density

$$A_{sine}(f_n) = \sqrt{\pi f_n W_{\ddot{U}}(f_n)}. \tag{9.2}$$

J. Wijker, *Miles' Equation in Random Vibrations*, Solid Mechanics and Its Applications 248, https://doi.org/10.1007/978-3-319-73114-8_9

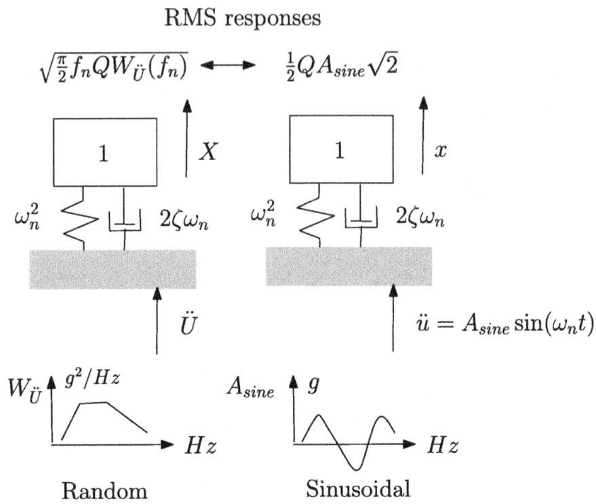

Fig. 9.1 Equivalence between random and sinusoidal base excitation of two SDOF systems

9.2 Sinusoidal-Random Equivalence of Responses

A number of publications about the equivalence between sinusoidal and random vibration is discussed, e.g., [1, 2].

In [3], it is suggested that a sinusoidal dwell at the resonance frequency may produce the same fatigue damage as random excitation if its r.m.s. response is about

$$G_{rms} = 1.5\sqrt{\frac{\pi f_n W_{\ddot{U}}(f_n)}{2Q}}. \tag{9.3}$$

The corresponding equivalent sinusoidal amplitude is about

$$A_{sine}(f_n) = 2.7\sqrt{\frac{f_n W_{\ddot{U}}(f_n)}{Q}}. \tag{9.4}$$

In [2], the sine sweep or dwell response equivalence is given by

$$A_{sine}(f_n) = 3.668\sqrt{\frac{\pi f_n W_{\ddot{U}}(f_n)}{2Q}}. \tag{9.5}$$

Another suggestion to calculate the equivalence between random and sinusoidal test is given in [3], which is dependent on the time duration of the random test t_o, the sweep rate during the sine vibration test $S = df/dt$, and the number of sweeps N through the resonance frequency during the sinusoidal vibration test. Taking into account fatigue damage the sinusoidal amplitude can be obtained by

$$A_{sine}(f_n) = 1.5\sqrt{\frac{2St_o W_{\ddot{U}}(f_n)}{N}}. \tag{9.6}$$

In case of logarithmic change of the frequency, e.g., n oct./min, the sweep rate S is given by $S(f_n) = \{n \ln(2)/60\} f_n$ [4], the number of sweeps is, in general, $N = 1$ and the time duration of a qualification random vibration test $t_o = 120$ s. (60 s for proto-flight and acceptance test).

Example

This example is based on an example discussed in [1]. The task is to convert a random vibration specification to a new sinusoidal vibration specification. The time duration of random acceleration specification is $t_{rnd} = 120$ s, and the PSD is given by $W_{\ddot{U}} = 0.02\,\mathrm{g}^2/\mathrm{Hz}$ from 20 Hz to 2000 Hz. The finite element analysis demonstrated a first fundamental vibration at a natural frequency $f_n = 90$ Hz and the amplification factor is $Q = 10$. The equivalent sinusoidal amplitude is

$$A_{sine} = \sqrt{\pi f_n W_{\ddot{U}}} = 2.3780\ \mathrm{g}.$$

The number sinusoidal cycles is given by

$$N_{sine} = t_{rnd} f_n = 10800 \text{ cycles}.$$

The sweep rate n (Oct/min) will be defined by the duration t_{rnd} of the sweep through the half power bandwidth $\Delta f_n = f_n/Q$, [4].

$$n = \frac{60 \ln\left(\frac{f_n + f_n/2Q}{f_n - f_n/2Q}\right)}{t_{rnd} \ln 2} = 0.0722 \text{ Oct/min}.$$

The number of cycles is

$$N = \frac{60}{n \ln 2} \frac{f_n}{Q} = 10800.$$

9.3 Chapter Summary

In general, the sinusoidal vibration (sweep) test is in the frequency range between 5 and 100 Hz and the random vibration test has a frequency range between 20–2000 Hz. Therefore, it is not so straightforward to replace a random vibration test by a equivalent sinusoidal vibration test or vice versa. Damaging and fatigue aspects have to be considered transferring a specific sinusoidal specification into equivalent random vibration specification; Chap. 10, in particular the section about the fatigue damage spectrum [5], may be very helpful to do so.

Table 9.1 Random vibration specification

Frequency band (Hz)	Spectral density (g^2/Hz)
20–50	0.01125
50–100	3 dB/oct
100	0.0225
200–500	4 dB/oct
500	0.05625
500–1000	−4 dB/oct
1000	0.0225
1000–2000	−3 dB/oct
G_{rms}	7.5 g

Problems

9.1 Prove that $W_{\ddot{U}} = \frac{A_{sine}^2}{\pi f_n}$.
Hint: Assume $Q = 1$.

9.2 This problem is based on an early version of the Soyuz L/V manual, Issue 1, revision 0, June 2006, Sect. 4.3.3.3, [6]. The verification of the S/C structure compliance with the random vibration environment in the 20–100 Hz frequency range shall be performed with one of the three following methodologies:

1. Perform a dedicated random vibration test.
2. Conduct the sine vibration qualification test up to 100 Hz and apply input levels high enough to cover the random vibration environment (equivalency obtained with Miles' equation)

$$G_{rms} = \sqrt{\frac{\pi}{2} f_n Q W_{\ddot{U}}(f_n)}.$$

3. Conduct the sine vibration qualification test up to 100 Hz so as to restitute the structural transfer function and demonstrate the compliance of the S/C secondary structure with the random vibration environment by analysis.

Above 100 Hz, S/C qualification with respect to random vibration environment is obtained through the acoustic vibration test.

In Fig. 9.2 is shown that an scientific instrument with mass m_s and stiffness k_s is mounted on top of the S/C with masses m_1, m_2 (kg) and spring stiffnesses k_1, k_2 (N/m). The instrument mass $m_s = 20$ kg and the lowest natural frequency is $f_s = 125$ Hz. Define the spring stiffness k_s.

The spring stiffness $k_1 = 10^7$ N/m and spring stiffness $k_2 = 2k_1$ N/m. The mass $m_1 = 100$ kg and mass $m_2 = 2m_1$ kg and the damping ratio $\zeta = 5\%$.

1. Perform the random response analysis with the random enforced vibration levels (20–2000 Hz) as provided in Table 9.1 using your favorite FEA software package.

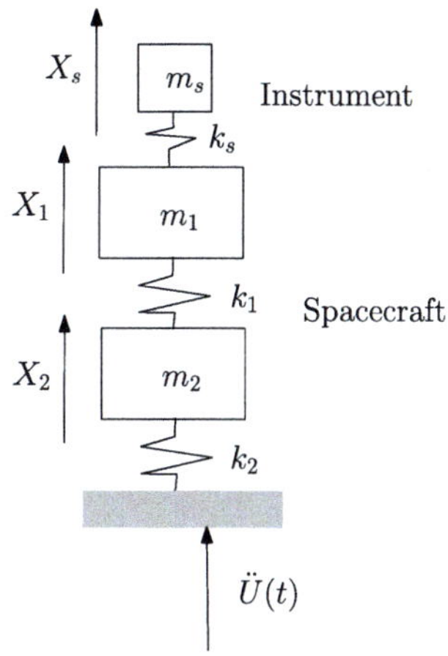

Fig. 9.2 Instrument mounted on top of S/C

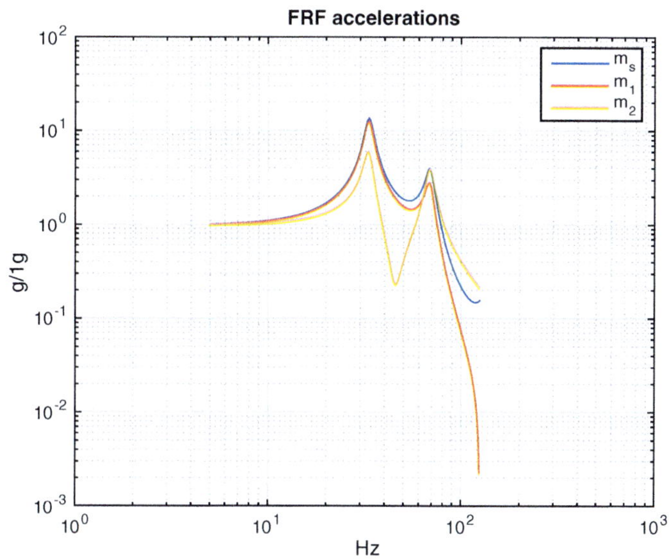

Fig. 9.3 Frequency response (transfer) functions S/C

- Calculate the undamped natural frequencies f_n (Hz) and associated modal effective masses M_{eff} (kg).
- Calculate r.m.s. acceleration of the masses m_s, m_1, m_2, $\ddot{x}_{s,rms}$, $\ddot{x}_{1,rms}$, $\ddot{x}_{2,rms}$, Number of frequency steps $N = 2000$.
- Discuss why the first vibration mode is most important when a base excitation response analysis is performed?

2. Calculate the equivalent sine excitation amplitude A_{sine} (g) (5–100Hz) using the following expression:
$$A_{sine} = \sqrt{2}\frac{G_{rms}}{Q}\ \text{g}.$$
Derive this equation and why do we use the $G_{rms} = \ddot{x}_{s,rms}$.
Perform a sinusoidal enforced acceleration with A_{sine} as amplitude and define maximum acceleration response $\ddot{x}_{s,rms}$ of mass m_s and associated excitation frequency. Discuss if we meet the requirement of the sinusoidal responses cover the random responses.
3. Calculate the frequency response (transfer) functions of the S/C between 5–100 Hz. (Fig. 9.3), Number of frequency steps $N = 200$. After that compute with the aid of Miles'equation the r.m.s. acceleration of the instrument m_s, using the natural frequency f_n of the fundamental mode and associated frequency transfer function $|H(f_n)|$ and PSD $W_{\ddot{U}}(f_n)$. Discuss this result compared to the methodologies 1 and 2.

Answers: $k_s = 1.2337e + 07$ N/m.

1. $f_n = (33.24, 68.63, 138.81)$ Hz, $M_{eff} = 277.56, 42.4, 0.01$ kg. $(\ddot{x}_{s,rms}, \ddot{x}_{1,rms}, \ddot{x}_{2,rms}) = 3.68, 3.27, 2.16$ g.
2. $A_{sine} = 0.52$ g. $\ddot{x}_{s,max} = 7.07$ g, $f_{max} = 32.87$ Hz.
3. $\ddot{x}_{s,rms} = 4.81$ g.

References

1. Wang FF, Crane E (2003) Relating sinusoid to random vibration for electronic equipment testing. COTS J, 6 p
2. Jayahari L, Praveen G (2005) Correlation of sinusoidal sweep test to field random vibrations, bth-amt-ex-2005/d-13-se. Master's thesis, Blekinge Institute of Technology, Karlskona, Sweden, April 2005
3. Harris CM, Crede CE (1976) Shock and vibration handbook, 2nd edn. McGraw-Hill Book Company, Maidenheach. ISBN 0-07-026799-5
4. Wijker JJ (2008) Spacecraft structures. Springer, Berlin. ISBN 978-3-540-755524
5. McNeill SI (2008) Implementing the fatigue damage spectrum and fatigue equivalent vibration testing. Sound and vibration, 79th Shock and vibration symposium, Orlando, Florida, 26–30 Oct 2008
6. Arianespace (2006) Soyuz from guiana space centre user's manual, issue 1, revision 0 edition, www.arianespace.com

Chapter 10
Characterisation and Synthesis of Random Acceleration Vibration Specifications

Abstract A number of damage spectra to characterise measured or computed random acceleration vibration spectra are discussed in this chapter. The characterisation is based on equivalent damage caused by extreme peaks (SRS, ERS, VRS) and Rayleigh distribution of peaks or cumulative damage (FDS), using relative displacements and pseudo-velocities. The response spectra are all based on the response of single degree of freedom (SDOF) systems exited to random accelerations, both in the time and frequency domain. The principles to compute the response spectra are illustrated and discussed. One general practical example is discussed in very detail. Miles' equation fulfills a key role in the synthesis process to generate equivalent random acceleration vibration specifications. An envelope of the damage response spectrum is achieved by dividing the spectrum into a number of fields or regions. The lower the number of field more severe and smoother equivalent random acceleration vibration specification is obtained.

Keywords Synthesis of acceleration spectra · Damage spectra · Equivalent spectra

10.1 Introduction

Subsystems (instruments, equipment, boxes) are to be qualified against rather severe mechanical random acceleration vibration test specifications as discussed in [1]. The random acceleration vibration test specifications are, in general, enforced accelerations at the interface between spacecraft and subsystems. The random vibrations are mainly induced by the acoustic loads exposed to the spacecraft during launch and performing acoustic tests, representing the launch environment. The acoustic loads (sound pressures) are assumed to be diffuse and are simulated in a reverberant chamber, like the Large European Acoustic Facility (LEAF) at ESA/ESTEC, Noordwijk, The Netherlands. This chapter is based on [2].

The measured random accelerations or similar predictions are broad-banded and show many peaks as shown in Fig. 10.1 (blue line). These random acceleration measurements and predictions are converted into more or less equivalent smooth random acceleration vibration test specification, which represent as good as possible the

J. Wijker, *Miles' Equation in Random Vibrations*, Solid Mechanics and Its Applications 248, https://doi.org/10.1007/978-3-319-73114-8_10

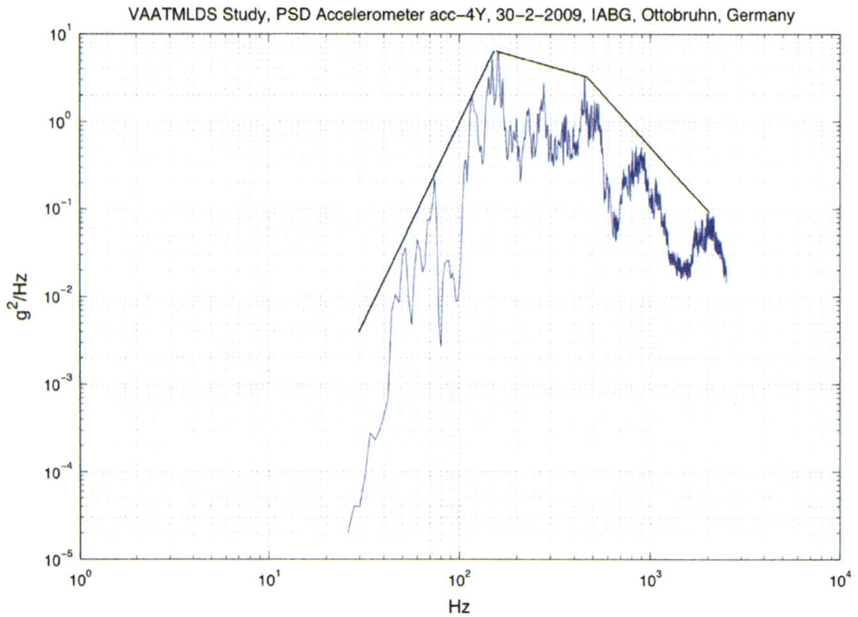

Fig. 10.1 Schematic random acceleration vibration test specification (black) representing measured PSD responses (blue) g^2/Hz

underlying measured and calculated random acceleration responses. The equivalent random acceleration vibration test specification shall not lead to under-testing or significant over-testing of the test item. An enveloping test random acceleration vibration specification is illustrated in Fig. 10.1 (black line), which is such severe that over-testing of the test item will certainly occur.

Several methods are available to reconstruct and characterise in a very structured manner the equivalent random acceleration vibration test specification from the measured and predicted random response data. The following characterisation methods are discussed in this chapter:

- Shock response spectrum (SRS) [3], extreme response spectrum (ERS) [4]: Both response spectra are based on extreme responses of SDOF systems excited by enforced random accelerations, the first in the time and the second in frequency domain.
- Vibration response spectrum (VRS) [5]: The VRS is the response spectrum based on the 3σ responses of series of SDOFs, with varying natural frequency, excited at the base by random acceleration in the frequency domain.
- Fatigue damage spectrum (FDS) [4]: This damage response spectrum represents the cumulative damage due to the random responses of series of SDOF systems, with varying natural frequency, excited by random acceleration, both in the time and frequency domain.

Another method to synthesise the equivalent random vibration spectra is a method applied to pseudo-stationary random vibration [6, 7] and is based on Rayleigh distribution of peaks (damage potential) [8].

Miles' equation [9] is in the synthesis process of equivalent random acceleration vibration specifications from the different response spectra of great importance.

10.2 Previous Work

The SRS was already mentioned by M.A. Biot in 1933 [10] and later in 1941 [11]. The theoretical description of the SRS was done within the frame of earthquake engineering. A mechanical analyzer was developed to predict stresses in structural systems under transient impulse. The concept of SRS of accelerations is nowadays still in use in spacecraft structure engineering to characterize the severity of high-frequency transient accelerations, such as separation of launch vehicle stages, shroud and the separation of spacecraft [12]. In SVM-5 [3], the principles and methods to analyze shocks are discussed in depth. Smallwood developed in [13] a very efficient recursive formula to calculate the SRS. In [14, 15], the Smallwood recursive method (z-transform) was presented as a Matlab® or Octave® script. This script is applied to calculate the acceleration SRS in the time domain.

Lalanne discussed in his book "Specification Development" [4] the principle of the ERS and FDS for both sinusoidal and random vibration and the FDS for shocks as well.

Halfpenny et al. discussed in their paper [16] the application of the SRS, ERS for peak accelerations, and the FDS, to represent the cumulative fatigue damage caused by relative displacements (stresses), to describe new methods for vibration qualification of equipment on aircraft. Equivalent ERS and FDS are calculated for the proposed qualification random acceleration vibration test specification, which envelopes the flight ERS and SRS. The ERS is calculated in the frequency domain using numerical integration of the vibration spectrum and/or Miles' equation [9]. The peak values of stationary Gaussian process are discussed in detail in [17, 18]. In [16], the FDS is accurately obtained by numerical integration of the response spectrum or can be approximated applying Miles' equation. The approximate equation is used to calculate the equivalent random acceleration vibration spectrum.

Halfpenny described in his paper [19] the calculation of the FDS of random vibrations both in the time and frequency domain. The random acceleration vibration spectrum is synthesised into the time domain by the summation of sine waves and the associated frequency, which in turn is applied to calculate the FDS in the time domain. The rain-flow counting method is made available in Matlab® scripts; however, a number of rain-flow counting methods are discussed in [20]. The synthesised random vibration test spectrum is calculated from the FDS, both in time and frequency domain, using the Miles' approximation for the FDS. The FDS is based on the relative displacements of the SDOF systems base excited by the random accelerations.

McNeil described in his paper [7] the application of the FDS, however, not based on the relative displacement, but on the relative pseudo-velocities of the SDOF systems. He stated that at resonance the pseudo-velocity is roughly proportional to the stress. The FDS is calculated both in the frequency and time domain. In the time domain, rain-flow counting is used to extract amplitudes and associated number of cycles. The equivalent random vibration spectrum is calculated inverting the FDS based on Miles' approximation. The second method discussed in this paper is based upon the Rayleigh distribution of peaks for non-stationary random accelerations.

Eaton [21] described in his paper presented on the 2012 SCLV Conference a test tailoring methodology for equipment based on the ERS and FDS.

Dimaggio et al. [6] mentioned the Rayleigh distribution of peaks a damage-based approach. The synthesis of the equivalent random vibration spectrum is again based on Miles' equation. Irvine presented on the 2012 SCLV conference a paper [22] a comparison of the damage potential method with the SRS, ERS, and VRS methods.

The VRS is described by Irvine in [5], which is used to synthesise random vibration spectra by enveloping the VRS. Miles' equation is applied to obtain synthesised random vibration spectra. The VRS spectrum is very similar to the ERS.

In the previous referenced papers (e.g., [6]), it was recommended to vary the amplification factor and Basquin's exponent of the s-N curve, such that $Q = 10, 25, 50$ and $b = 4, 8, 12$. The worst synthesised random vibration spectra shall be applied to testing.

10.3 Random Vibration Spectra

In this section, a number of random vibration spectra will be explained and discussed.

10.3.1 Shock Response Spectrum (SRS)

The SRS was first introduced by Biot [10, 11] within the frame of earthquake engineering and has been in existence for a long time.

The SRS is a graphical representation of an arbitrary transient acceleration input, how a single degree of freedom system (SDOF) (mass-spring-damper) responds to that input. Actually, it shows the peak acceleration response in principle of an infinite number of SDOF systems, each of which has different natural frequencies ($\omega_i = 2\pi f_i,\ i = 1, 2, \ldots$). This is illustrated in Fig. 10.2, where the SDOF systems are mass normalised. In fact, the SRS analysis is the maximum response of a series of SDOF systems having the same damping to a given transient signal. In practice, a damping ratio $\zeta = 0.05$ ($Q = 10$) is assumed.

The analysis procedure to compute the SRS is described in [14, 15], where a MATLAB® script has been presented based on the recursive formula proposed by Smallwood [13].

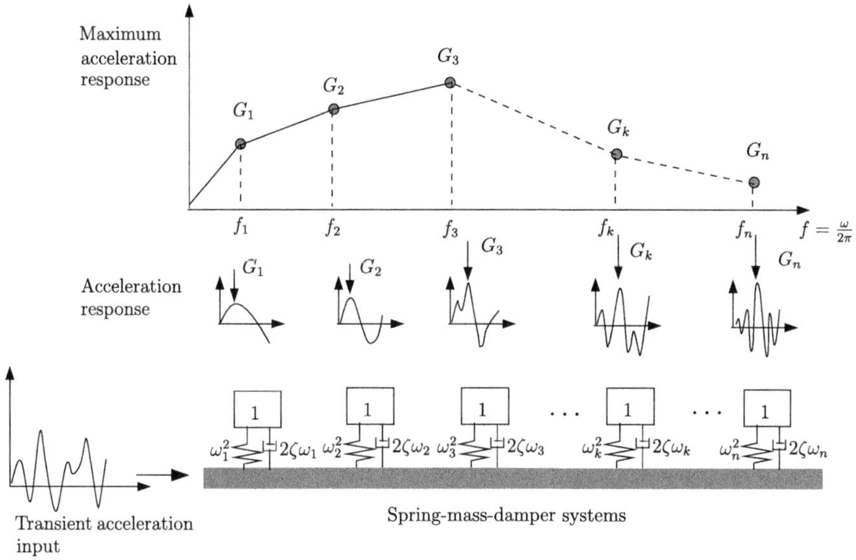

Fig. 10.2 How a shock response spectrum is developed

10.3.2 Vibration Response Spectrum (VRS)

Tom Irvine introduced the VRS in [5], which is similar to the SRS; however, the transient acceleration input is now replaced by random acceleration PSD input $W_{\ddot{u}}(f)$ (g^2/Hz), and the transient maximum responses are replaced by the 1σ absolute responses of the SDOF systems. In general, the damping ratio is taken $\zeta = 0.05$ ($Q = 10$) and is the same for all SDOF systems.

The graphical representation of the VRS is illustrated in Fig. 10.3. The SDOF systems are mass normalised. The standard deviation of the response of the SDOF system enforced at the base by random acceleration input can be calculated by the following expression

$$\sigma_i = \sqrt{\int_0^{\infty} |H_i(f)|^2 W_{\ddot{u}}(f) df} \approx \sqrt{\sum_{f_{min}}^{f_{max}} |H_i(f)|^2 W_{\ddot{u}}(f) \Delta f}, \; i = 1, 2, \ldots, N, \tag{10.1}$$

where the frequency transfer function (FRF) $H_i(f)$ is given by

$$H_i(f) = \frac{1 + 2j\zeta f/f_i}{1 - (f/f_i)^2 + 2j\zeta f/f_i}, \tag{10.2}$$

and $f_i = \omega_i/2\pi$ is the natural frequency.

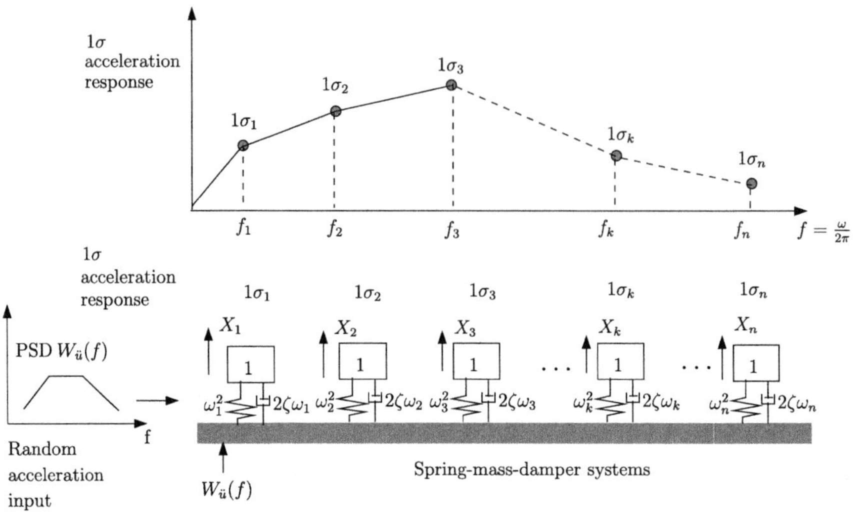

Fig. 10.3 How a vibration response spectrum is developed

If the input PSD $W_{\ddot{u}}(f)$ is relatively flat at frequencies near the natural frequency f_i, Eq. (10.15) can be approximated by Miles' equation [23]:

$$\sigma_i \cong \sqrt{\frac{\pi f_i W_{\ddot{u}}(f_i)(1+4\zeta^2)}{4\zeta}} \approx \sqrt{\frac{\pi f_i Q W_{\ddot{u}}(f_i)}{2}},\ i = 1, 2, \ldots, N. \tag{10.3}$$

The VRS is defined as

$$VRS(f_i) = 3\sigma_i,\ i = 1, 2, \ldots, N. \tag{10.4}$$

The VRS equivalent random acceleration vibration specification $W_{\ddot{u}}(f_i)$ for the enforced acceleration can be calculated by inverting of (10.3)

$$W_{VRS,\ddot{u}}(f_i) = \frac{2(VRS)^2(f_i)}{9\pi f_i Q},\ i = 1, 2, \ldots, N. \tag{10.5}$$

10.3.3 Extreme (Peak) Response Spectrum (ERS)

The ERS is similar to the SRS because peak values of the SDOF random response are depicted graphically in the ERS. Lalanne already discussed in [4] the ERS for random vibrations. The $1\sigma_i$ response of the SDOF system "i" is multiplied by a random peak factor C_i. The expected value of the peak factor of the random response of the SDOF system is given by [18]:

$$E[C_i] = \sqrt{2\ln(f_i T)} + \frac{\gamma}{\sqrt{2\ln(f_i T)}}, \quad (10.6)$$

and the standard deviation of the peak factor is

$$\sigma[C_i] = \frac{\pi}{\sqrt{6}} \frac{1}{\sqrt{2\ln(f_i T)}}, \quad (10.7)$$

where f_i is the natural frequency of the SDOF system, T the time duration of the random accelerations process, and $\gamma = 0.5772$ is the Euler constant [24]. Further, it is assumed that $f_i T \gg 1$, and

- the random response X_i of the SDOF system has a Gaussian distribution, and
- the peak values of the random response of the SDOF system are statically independent. The statistically independence assumption is acceptable if $E[C_1]/\sigma[C_1] > 3.5$ and $f_i T > 250$ for a Gaussian process [25].

With a given standard deviation σ_i of the random response of the SDOF system, the maximum expected extreme (peak) acceleration response spectrum $ERS_a(f_i)$ can be calculated using:

$$ERS_a(f_i) = E[C_i]\sigma_i, \; i = 1, 2, \ldots, N. \quad (10.8)$$

The expected peak values for the displacement, the $ERS_d(f_i)$ displacement spectrum, is given by [16]

$$ERS_d(f_i) = 9.81^2 \frac{E[C_i]\sigma_i}{\omega_i^2}, \; i = 1, 2, \ldots, N, \quad (10.9)$$

in case the PSD input spectrum $W_{\ddot{u}}(f)$ is specified in g^2/Hz.

The ERS equivalent random acceleration vibration specification $W_{ERS,\ddot{u}}(f_i)$ for the enforced acceleration can be calculated with the aid of inverting Eq. (10.8)

$$W_{ERS,\ddot{u}}(f_i) = \frac{2ERS^2(f_i)}{E(C_i)^2 \pi f_i Q}, \; i = 1, 2, \ldots, N. \quad (10.10)$$

Because the SRS is also based on peak responses, the equivalent random acceleration vibration specification can be obtained using Eq. (10.10),

$$W_{SRS,\ddot{u}}(f_i) = \frac{2SRS^2(f_i)}{E(C_i)^2 \pi f_i Q}, \; i = 1, 2, \ldots, N. \quad (10.11)$$

10.3.4 Fatigue Damage Spectrum (FDS)

In this section, the fatigue damage spectrum will be based on the relative displacements FDS_d and pseudo-velocities [7] FDS_{pv}.

10.3.4.1 FDS_d in Frequency Domain

The expected cumulative fatigue damage $E[D(T)]$ (in one SDOF system) involves the Palgren–Miner fatigue accumulation rule [26, 27] in combination of the s-N curve $N_k s_k^b = C$ (N_k is the number of allowable cycles at stress level s_k) for a narrow-banded Gaussian process the cumulative damage and is given by [8, 28] for $f_1,\ i = 1, 2, \dots N$:

$$E[D(T, f_i)] = \frac{f_i T}{C}(\sqrt{2}\sigma_s)^b \Gamma\left(1 + \frac{b}{2}\right), \tag{10.12}$$

where C is a material constant (Basquin coefficient), b is the fatigue exponent (Basquin's exponent), f_i is the natural frequency of the SDOF system, T the time duration, σ_s is the standard deviation of the stress, and Γ is the gamma function.

The stress s is proportional to the relative displacement of the SDOF multiplied by a constant K. The standard deviation of the stress $\sigma_{s,i}$ can be obtained as follows:

$$\sigma_{s,i} = K\frac{9.81}{(2\pi f_i)^2}\sqrt{\frac{\pi}{2} f_i Q W_{\ddot{u}}(f_i)}, \tag{10.13}$$

because the PSD spectrum $W_{\ddot{u}}$ of the enforced random acceleration is most times specified in g^2/Hz.

The FDS_d corresponding to the natural frequency of the SDOF system "i" is now defined for $f_i,\ i = 1, 2, \dots N$:

$$FDS_d(T, f_i) = f_i T\frac{K^b}{C}\left[\frac{9.81^2 Q W_{\ddot{u}}(f_i)}{2(2\pi f_i)^3}\right]^{\frac{b}{2}} \Gamma\left(1 + \frac{b}{2}\right). \tag{10.14}$$

In general, the constants are taken as $K = C = 1$, the exponent $b = 4, 8, 12$, and the amplification factor $Q = 10, 25, 50$.

Instead of using Miles' equation, the standard deviation of the relative displacement response $\sigma_{d,i}$ of the SDOF system can be computed as follows

$$\sigma_{d,i} = \sqrt{\int_0^\infty |H_{d,i}(f)|^2 W_{\ddot{u}}(f)df} \approx \sqrt{\sum_{f_{min}}^{f_{max}} |H_{d,i}(f)|^2 W_{\ddot{u}}(f)\Delta f},\ i = 1, 2, \dots, n, \tag{10.15}$$

where the frequency transfer function (FRF) $H_i(f)$ is given by

$$H_{d,i}(f) = \frac{1}{(2\pi f_i)^2} \frac{1}{(1 - (f/f_i)^2 + 2j\zeta f/f_i)}. \tag{10.16}$$

The standard deviation of the stress s_i is $\sigma_{s_i} = 9.81K\sigma_{d,i}$. The mean value of the stress is assumed to be zero.

The synthesised equivalent random acceleration vibration spectrum $W_{FDS_d,\ddot{u}}(f_i)$ can be calculated by the inverse of (10.14)

$$W_{FDS_d,\ddot{u}}(f_i) = \frac{2(2\pi f_i)^3}{9.81^2 Q} \left[\frac{C\, FDS_d(f_i)}{K^b f_i T \Gamma\left(1 + \frac{b}{2}\right)} \right]^{\frac{2}{b}}, \; i = 1, 2, \ldots, N. \tag{10.17}$$

10.3.4.2 FDS$_d$ in Time Domain

If the random acceleration is provided in the time domain or is synthesised into the time domain, the cumulative damage D is given by the Palgren–Miner rule, which tells us that cumulative damage D is

$$D = \sum_{k=1}^{N} \frac{n_k}{N_k}, \tag{10.18}$$

where n_k is the number of stress oscillations at stress level s_k, and N_k is the number of allowable oscillation at stress level s_k given by the s-N curve

$$s_k^b N(s_k) = C. \tag{10.19}$$

The generation of the FDS$_d$ in the time domain is symbolically illustrated in Fig. 10.4. The random enforced transient acceleration is applied to the base of each of the SDOF systems. From the calculated random relative displacements $z(t)$, per SDOF system, the numbers of cycles and associated "stress" peaks are extracted using a rain-flow counting procedure [29, 30]. Using "stress" peaks and cycles, the fatigue damage spectrum can be obtained. Again, the stress constant and the Basquin's coefficient are assumed to be $K = C = 1$, and the Basquin exponent may vary $b = 4, 8, 12$.

10.3.4.3 FDS$_{pv}$ in Frequency Domain (Spectral Method)

In [7], the pseudo-velocity $pv(t)$ of the SDOF system is applied to calculate the FDS$_{pv}$, because the pseudo-velocity is roughly proportional to stress for many structures; a scale factor exists between the stress σ and the pseudo-velocity $pv, \sigma = Kpv$,

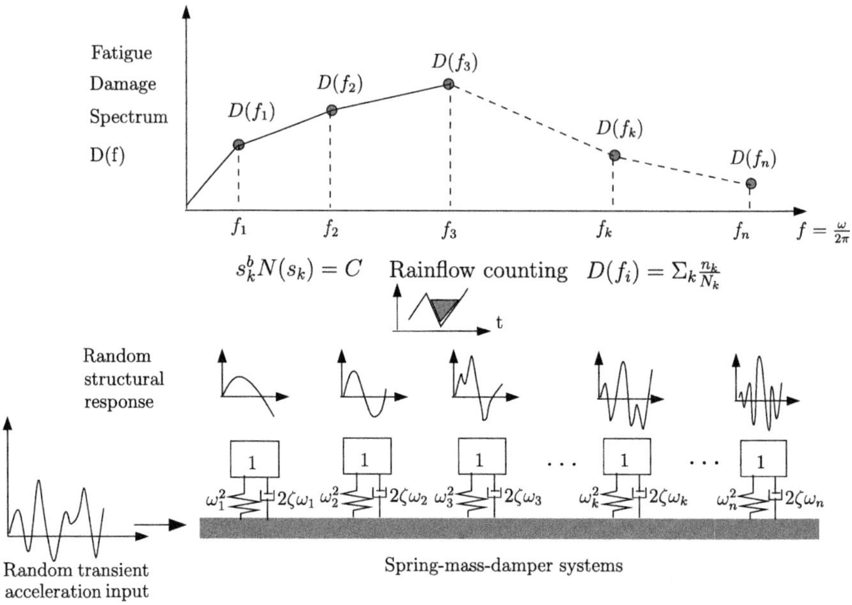

Fig. 10.4 How a time domain damage fatigue spectrum is developed

[31, 32]. The maximum strain amplitude ε is proportional to the "vibration Mach number" v/c, where v is the maximum vibratory velocity, and c is the speed of sound in the material. The maximum strain is given by $\varepsilon = k(v/c)$, where $k = 0.145 - 2.00$ for many different configurations [31].

The pseudo-velocity is defined as the relative displacement response $z(t)$ of the SDOF system multiplied by the natural frequency of that SDOF, $pv(t) = z(t)(2\pi f_i)$.

The expected cumulative fatigue damage $E[D(T)]$ involves the Palgren–Miner fatigue accumulation rule [27] in combination of the s-N curve $N_k s_k^b = C$ (N_k is the number of allowable cycles at stress level s_k) for a narrow-banded Gaussian process the expected cumulative damage of FDS and is given by [8, 28], for f_1, $i = 1, 2, \ldots N$

$$FDS_{pv}(f_i) = E[D(T, f_i)] = \frac{f_i T}{C}(\sqrt{2}\sigma_s)^b \Gamma\left(1 + \frac{b}{2}\right) = \frac{f_i T}{C} K^b (2\sigma_{pv}^2)^{\frac{b}{2}} \Gamma\left(1 + \frac{b}{2}\right). \tag{10.20}$$

The standard deviation of pseudo-velocity $\sigma_{pv,i}$ is expressed as follows:

$$\sigma_{pv,i} = 9.81\sqrt{\frac{1}{8\pi f_i} Q W_{\ddot{u}}(f_i)}, \tag{10.21}$$

where the PSD function $W_{\ddot{u}}$ is most times given in g^2/Hz.

In general, the constants are taken as $K = C = 1$, the exponent $b = 4, 8, 12$, and the amplification factor $Q = 10, 25, 50$.

Instead of using Miles' equation, the standard deviation of the pseudo-velocity response $\sigma_{pv,i}$ of the SDOF system can be numerically calculated as follows

$$\sigma_{pv,i} = 9.81\sqrt{\int_0^{\infty} |H_{pv,i}(f)|^2 W_{\ddot{u}}(f)df} \approx \sqrt{\sum_{f_{min}}^{f_{max}} |H_{pv,i}(f)|^2 W_{\ddot{u}}(f)\Delta f}, \; i = 1, 2, \ldots, n, \tag{10.22}$$

where the frequency transfer function (FRF) $H_{pv,i}(f)$ is given by

$$H_{pv,i}(f) = \frac{1}{(2\pi f_i)} \frac{1}{(1 - (f/f_i)^2 + 2j\varsigma f/f_i)}. \tag{10.23}$$

The standard deviation of the pseudo-velocity s_i is $\sigma_{s_i} = K\sigma_{pv,i}$. The mean value of the stress is zero.

The PSD values $W_{FDS_{pv}\ddot{u}}(f_i), i = 1, 2, \ldots, N$ of the random acceleration vibration specification can be obtained inverting Eq. (10.20) in combination with (10.21). Thus, we get

$$\sigma_{pv}^2(f_i) = \frac{1}{2}\left(\frac{C \; FDS_{pv}(f_i)}{f_i T K^b \Gamma(1 + \frac{b}{2})}\right)^{\frac{2}{b}}, \tag{10.24}$$

and

$$W_{FDS_{pv}\ddot{u}}(f_i) = \frac{8\pi f_i \sigma_{pv}^2(f_i)}{9.81^2 Q}, \; i = 1, 2, \ldots, N. \tag{10.25}$$

10.3.4.4 FDS$_{pv}$ in Time Domain

The calculation of FDS$_{pv}$ is identical to calculation of the FDS$_d$; however, the stress is now proportional to the pseudo-velocity. If the random pseudo-velocity is provided in the time domain, the cumulative damage D is given by the Palgren–Miner rule, which tells us that cumulative damage D is

$$D = \sum_{k=1}^{N} \frac{n_k}{N_k} = \frac{K^b}{C} \sum_{k=1}^{N} n_k pv_k^b, \tag{10.26}$$

where n_k is the number oscillations at pseudo-velocity level pv_k, and N_k is the number of allowable oscillation at pseudo-level level pv_k given by the $s - N$ curve

$$s_k^b N(s_k) = K^b pv^b N(pv) = C. \tag{10.27}$$

The constants are taken $K = C = 1$ and, in general, the exponent $b = 4, 8, 12$.

The calculation procedure for FDS_{pv} is the similar to the calculation of FDS_d as shown in Fig. 10.4.

10.3.5 Pseudo-stationary Random Vibration, Damage Potential

The method of pseudo-stationary random vibration described in this section was discussed in [6, 7]; however, McNeil applied this method using pseudo-velocities, as discussed in Sects. 10.3.4.3 and 10.3.4.4.

The acceleration response of a lightly damped SDOF system excited by a zero-mean, stationary Gaussian white noise excitation is narrow-banded, and the probability of peak values is a Rayleigh distribution function and given by

$$f(A) = \left(\frac{A}{\sigma^2}\right)\exp\left(\frac{-A^2}{2\sigma^2}\right), \quad A \geq 0, \tag{10.28}$$

where A is the amplitude, and σ is the standard deviation. For a zero-mean response, the standard deviation is equal to the r.m.s. value of the response. The probability the amplitude is greater than A, $A >$ can be obtained by integrating the probability density function from A to ∞

$$P(A >) = \int_A^\infty f(a)da = \exp\left(\frac{-A^2}{2\sigma^2}\right). \tag{10.29}$$

If we consider a stationary test of duration T_o, the total time $T(A >)$, or cumulative duration, spent during which response cycles with amplitudes exceeding A occur is

$$T(A >) = T_o \exp\left(\frac{-A^2}{2\sigma^2}\right). \tag{10.30}$$

For a SDOF system, with natural frequency f_i, the mean square acceleration response to a broad-band Gaussian excitation with a power spectral density W (g^2/Hz) is approximated by Miles' equation

$$\sigma^2 = \frac{\pi}{2} f_i Q W(f_i). \tag{10.31}$$

Taking the natural logarithm of both sides of (10.30) and substituting (10.31) yields

$$\ln\left(\frac{T_o}{T(A >)}\right) = \frac{1}{\pi f_i W(f_i)} \frac{A^2}{Q}. \tag{10.32}$$

For a particular natural frequency f_i, the function (10.32) appears as a straight line on a plot of $\ln T(A >)$ versus linear A^2/Q as shown in Fig. 10.5. In Fig. 10.5, A_{max} is the expected maximum amplitude at the exceedence duration of one period $1/f_i$. Prescribing a test based on an enveloping Rayleigh line guarantees that the damage potential of the test, for a resonance of that frequency, envelops the damage potential of flight loads over the range of damping uncertainty. The Rayleigh line starts at T_o and $A^2/Q = 0$ and goes to a minimum time, the duration of the period $1/f_i$, associated with A^2_{max}/Q. The total time $T_o = n_{A_{min}>}/f_i$, where $n_{A_{min}>}$ the number of cycles with amplitudes $A \geq A_{min>}$. The number of cycles can be obtained by the rain-flow counting method. The equivalent power spectral density $W_{DP,\ddot{u}}(f_i)$ corresponding to the line T_o to A^2_{max}/Q can be calculating by the following expression

$$W_{DP,\ddot{u}}(f_i) = \frac{(A^2_{max}/Q)}{\pi f_i \ln(f_i T_o)}. \tag{10.33}$$

To construct the Rayleigh line (Fig. 10.5), the following step-by-step procedure is to be done:

1. Translate the power spectral density spectra of the random vibration into a time domain random excitation.
2. Determine the absolute random acceleration response of the SDOF system, with natural frequency f_1, excited by the random excitation in the time domain.
3. Perform a rain-flow counting to analyze the spectrum of amplitudes and corresponding number of cycles
4. Select a number of amplitudes $A_k,\ k = 1, 2, \ldots$, analyze the number of cycles $n(A_k >),\ k = 1, 2, \ldots$, and calculate the durations $T(A_k >) = n(A_k >)/f_i,\ k = 1, 2, \ldots$. The minimum number of cycles beyond A_{max} is $n(A_{max} >) = 1$, with $T(A_{min} >) = 1/f_i$, and the number of cycles beyond A_{min} is $n(A_{min} >= T_o/f_i$. This step is illustrated in Fig. 10.6.

10.4 Synthesis of Equivalent Random Acceleration Vibration Specification

It is assumed that the equivalent random acceleration vibration spectra represent the same amount of (fatigue) damage as expected for the original measured or calculated spectrum. The reconstruction or synthesis of an equivalent random acceleration vibration specification $W_{\ddot{u}}(f)$ can be done using the response spectra: SRS, VRS, ERS, FDS_d, FDS_{pv}, and the Rayleigh line (damage potential). The response spectra computing in the time domain can be used as well. For that purpose, the Eqs. (10.5), (10.10), (10.11), (10.17), (10.25), and (10.33) can be applied. Miles' equation has a key role computing the equivalent random acceleration specifications. The pro-

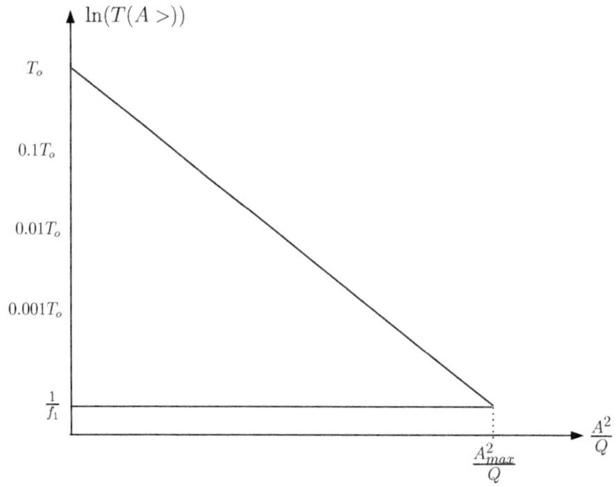

Fig. 10.5 Cumulative duration plot for Rayleigh distributed maxima

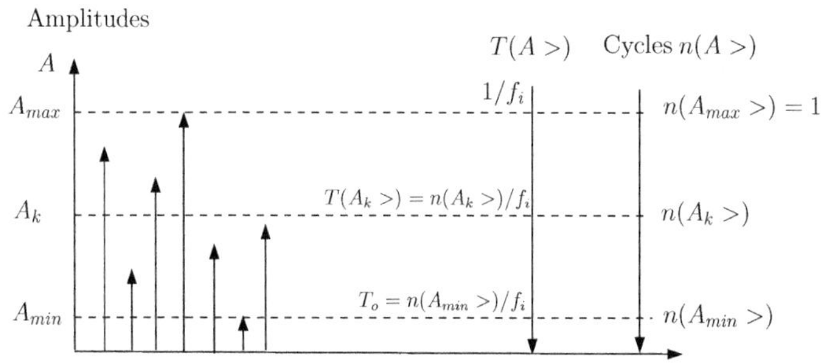

Fig. 10.6 The evaluation of $T(A >)$ versus the amplitude A

cedure to reconstruct the equivalent random acceleration vibration specification is illustrated in Fig. 10.7a, b. In order to address the variations in damping and s-N curve fatigue exponent b, the response spectra and Rayleigh line shall be computed with amplification factor values of $Q = 10, 25, 50$ and exponents $b = 4, 8, 12$ for each natural frequency f_i of the SDOF systems. The greatest power spectral density value of $W_{\ddot{u}}(f_i)$ over the 3–9 variations of Q and b is used as the random equivalent vibration specification level at frequency $f = f_i$. Finally, the complete random acceleration vibration specification is established. However, engineering judgement is still needed.

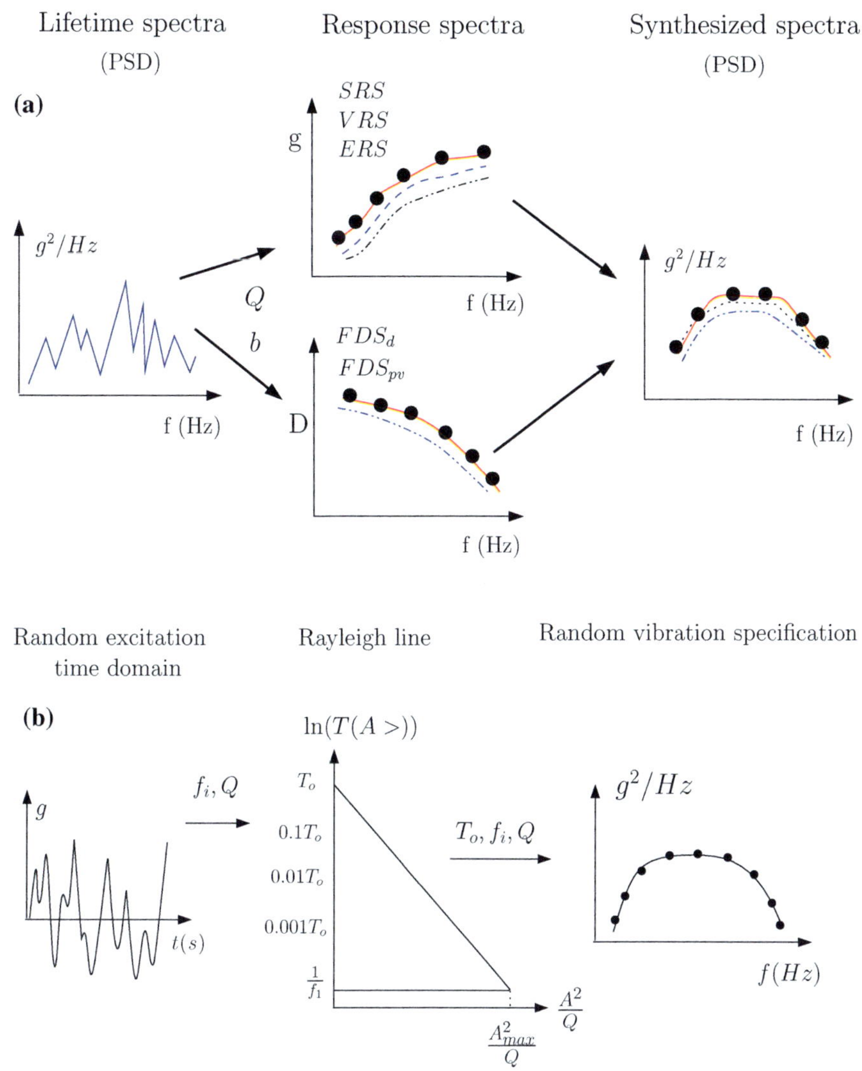

Fig. 10.7 Schemes to synthesize the random acceleration vibration specification

10.5 Application

10.5.1 Random Vibration Measurement During a Acoustic Test

Within the frame of the ESA/TRP study, "Vibro-Acoustic Analysis Test methods for Large Deployable Structures" (VAATMLDS) [33] acoustic tests were performed on

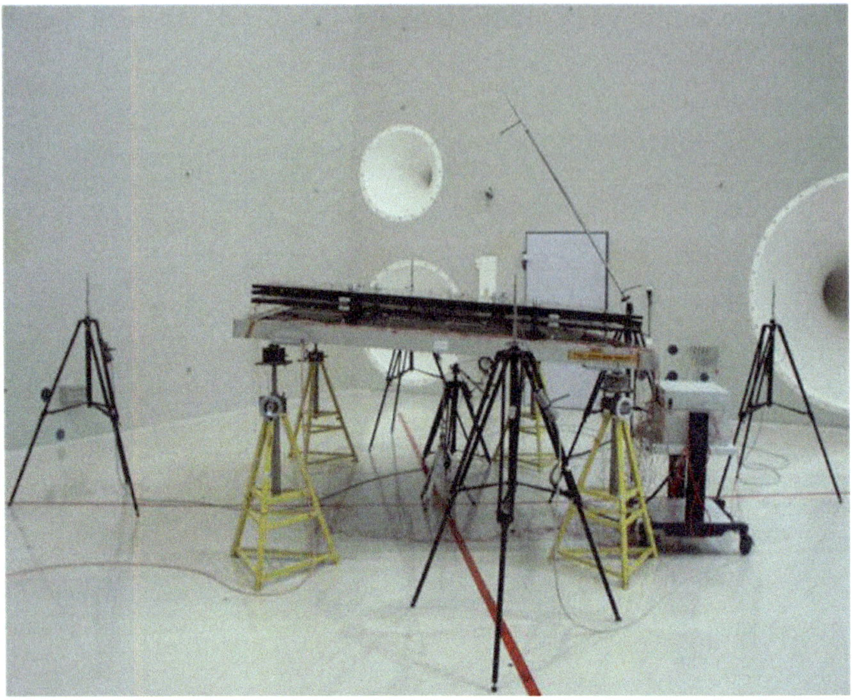

Fig. 10.8 VAATMLDS bread board 3 panel solar array wing in IABG acoustic chamber

the VAATMLDS bread board solar array wing in the acoustic reverberant chamber of IABG [34] (Fig. 10.8).

The read-out of accelerometer Acc-4Y (Fig. 10.9) is the random acceleration spectrum, for which a smooth and shaker controllable random vibration acceleration test spectrum should be generated, as illustrated in Fig. 10.1. The read-out of accelerometer Acc-4Y is in fact the raw random acceleration vibration specification. The number of data points is 1251, and the frequency increment $\Delta f = 2\,\text{Hz}$. The generation of the random acceleration vibration test specification will be performed in a more or less structured manner applying the response spectra: SRS, ERS, VRS, FDS and, the Rayleigh distribution of peaks. To account for uncertainties in damping and Palgren–Miner cumulative damage rule, the amplification factor (quality factor) will be varied $Q = 10, 25, 50$ and fatigue exponent $b = 4, 8, 12$. Matlab® is applied to perform all computations.

The random acceleration spectrum is synthesised into the time domain (waveform) using the method described in [35], in combination with the Fourier transform [36]. The synthesised equivalent signals in the time domain are shown Fig. 10.10. It should be noticed that the synthesised time domain accelerations are random and will vary from analysis to analysis, however, having the same mean and standard deviation. This is due to the random frequency shifts in the arguments of the sine waves. The time

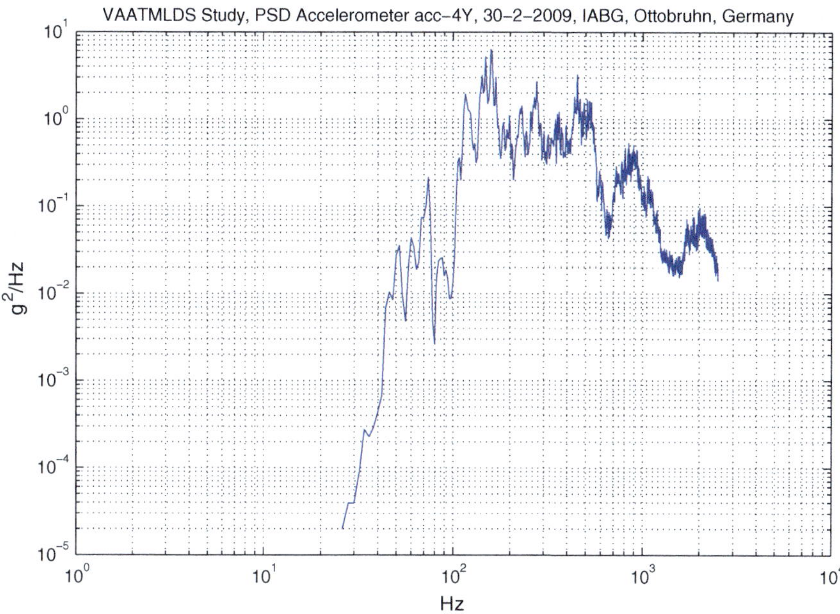

Fig. 10.9 PSD measurement of accelerometer 4-Y, Grms = 24.98 g

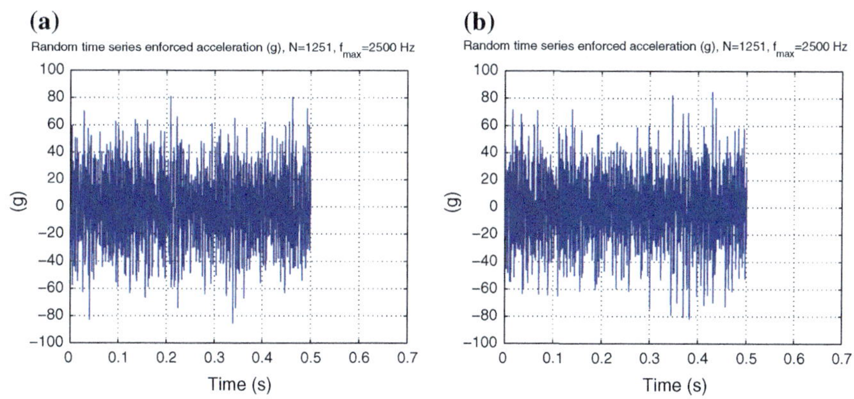

Fig. 10.10 Synthesised time domain random accelerations, $\mu = 0$, $1\sigma = 24.98\,\mathrm{g}$, $\lambda = 0.01$, $\gamma = 2.97$ (λ, γ mean values ten samples) [37]

increment $\delta t = 1/2f_{max} = 2.0 \times 10^{-4}\,\mathrm{s}$, and the total time $T = 2N\delta t = 0.5004\,\mathrm{s}$. The mean value of the all time domain signals is $\mu = 0$, and the standard deviation $1\sigma = r.m.s. = 24.98\,\mathrm{g}$. In case the time domain random acceleration is a Gaussian process, the skewness $\lambda = 0$ and the kurtosis $\gamma = 3$ [37]. A skewness $\gamma > 3$ will result in higher peaks, and a skewness $\gamma < 3$ will result in lower peaks compared to the ideal Gaussian process.

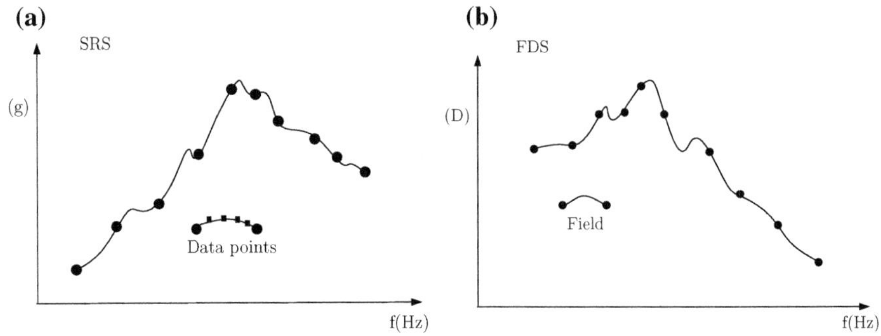

Fig. 10.11 Division of spectra in fields

The measured accelerations of accelerometer Acc-4Y, during the acoustic test, are now available both in the frequency and time domain. The response spectra: SRS, ERS, VRS, FDS_d, FDS_{pv}, and the distribution of Rayleigh peaks (damage potential) can now be computed.

The response spectra will be divided into N_{field} fields containing each n_{field} data points. This is illustrated in Fig. 10.11. The parameter N_{field} is set to $N_{field} = 100$. In each field, the maximum value of the spectrum, and corresponding frequency, is detected and used to calculate the synthesised PSD of the random acceleration vibration specification. The local maximum may be an internal point or at the boundaries of the field. The first and last data points of the complete spectrum are considered too. The synthesised PSDs at the detected points are connected to each other assuming the log–log scales for the spectra and corresponding frequencies, respectively. Internal PSD values can be obtained by the following expression

$$N = \log\left(\frac{W_2}{W_1}\right) / \log\left(\frac{f_2}{f_1}\right), \; W_i = W_1\left(\frac{f_i}{f_1}\right)^N = W_1\left(\frac{f_i}{f_1}\right)^{\frac{m}{3}}, \tag{10.34}$$

where W_1 and W_2 are the PSD values of the extreme points, f_1 and f_2 are the associated frequencies, and m is the slope of the spectrum in dB/octave. Index i indicates a point with frequency f_i in between the extreme points. Equation (10.34) is applied to calculate the area (mean square) under the synthesised equivalent PSD spectrum.

10.5.2 Synthesis of Random Acceleration Vibration Specification from SRS and ERS

In this section, the synthesised equivalent random acceleration vibration specification is calculated based on the SRS and ERS spectra. The SRS and ERS have 1251 and

Table 10.1 Synthesis random acceleration vibration specification from ERS and SRS

	Q		
	10	25	50
W_{ERS} r.m.s.	26.1	26.3	26.9
W_{SRS} r.m.s. μ	24.0	23.1	22.5
W_{SRS} r.m.s. σ	0.8	0.6	0.4
Acc-4Y	25.0 G_{rms}		

2481 data points, respectively. Both are ranging from 20–2500 Hz. The SRS and ERS are computed in accordance to the methods mentioned in Sects. 10.3.1 and 10.3.3. The VRS is common to the ERS, and therefore not considered. The computed ERS and SRS are shown Fig. 10.12a.

The SRS and ERS overlay each other very well. Instead of computing the SRS in the time domain, the ERS is a very good representative shock spectrum. Most times in spacecraft structures engineering, the 3σ VRS spectrum is used; however, it is recommended to use the ERS as the SRS for a given random acceleration vibration specification.

The SRS and ERS are divided into $N_{field} = 100$ fields. For each field, the maximum value of the SRS and ERS is detected and applied to synthesise the reduced number of data points of the SRS and the ERS into PSD of the random acceleration vibration specification using Eqs. (10.11) and (10.10), respectively. Both random acceleration vibration specifications are shown in Fig. 10.12b. Further, the synthesised random acceleration vibration specifications W_{SRS} and W_{ERS} are compared to the original response spectrum of accelerometer Acc-4Y in Fig. 10.12c. The PSD spectra W_{SRS} and W_{ERS} are computed varying the amplification factor $Q = 10, 25, 50$. The synthesised random accelerations are calculated from the original spectrum as shown in Fig. 10.9, although, mean and standard deviation remain the same, but the distribution of the peaks alters in time, due to random arguments in the sine waves representation of the original PSD spectrum. That means that the SRS and W_{SRS} are more or less random. Therefore the mean and standard deviation of W_{SRS} is calculated from ten samples. The PSD spectrum W_{ERS} will not change. The r.m.s. values of the PSD spectra W_{SRS} and W_{ERS} are given in Table 10.1. The synthesised r.m.s. values of PSD spectra of W_{SRS} underestimate the r.m.s. value of the original PSD spectrum, and the synthesised r.m.s. values of the PSD spectrum of W_{ERS} show a higher r.m.s. values compared to the original PSD spectrum of the accelerometer Acc-4Y.

A further simplification of the random acceleration vibration spectra W_{SRS} and W_{ERS} can be achieved dividing the SRS and ERS spectra into less fields; however, this will result in higher r.m.s. values of the equivalent random acceleration vibration specifications.

In Fig. 10.12d, the equivalent ERS from the equivalent W_{ERS} is computed for $Q = 10$ and compared to the original ERS.

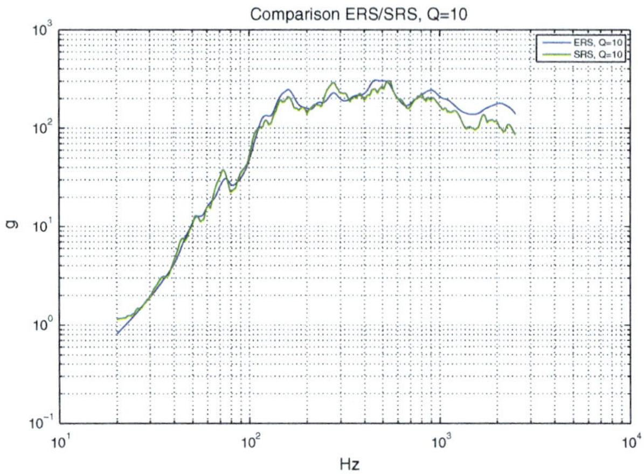

(a) SRS/ERS

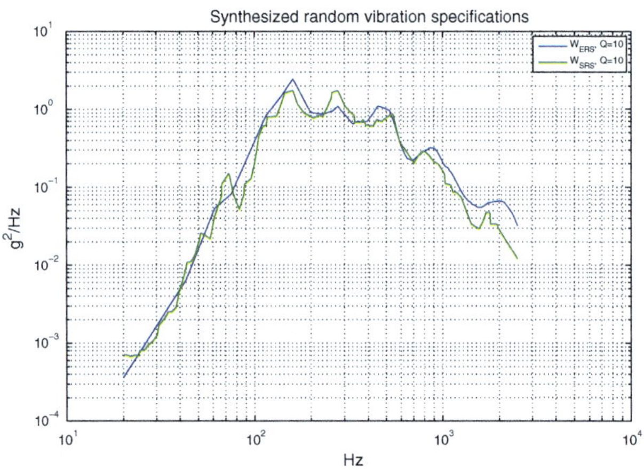

(b) W_{SRS}, W_{ERS}

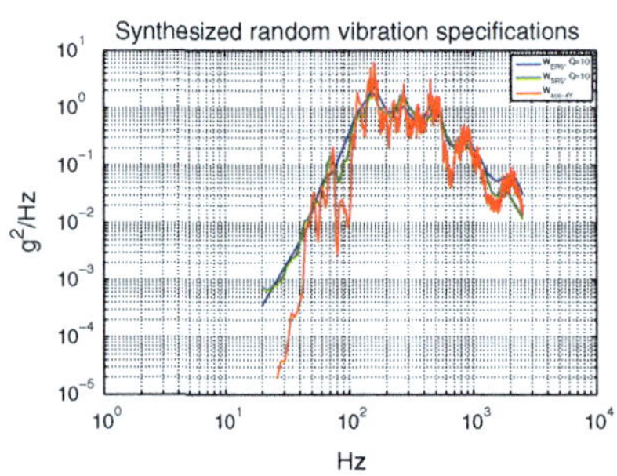

(c)

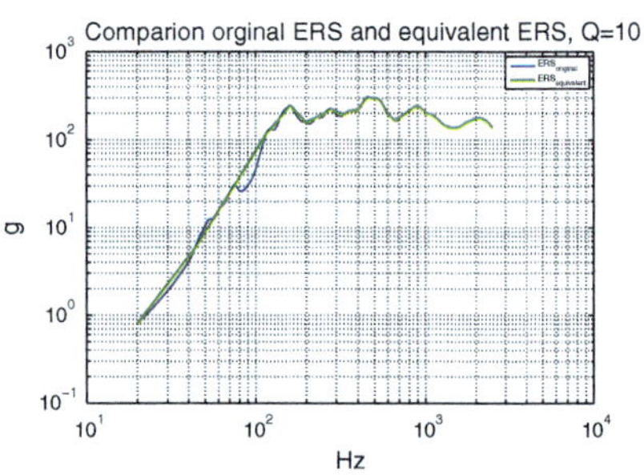

(d) ERS

Fig. 10.12 Synthesis of equivalent random acceleration vibration specification from SRS/ERS

Table 10.2 Synthesis random acceleration vibration specification from FDS_d

Q/r.m.s.	b					
	4		8		12	
	W_{Freq}	W_{RF} (RF*)	W_{Freq}	W_{RF}	W_{Freq}	W_{RF}
10	26.1	24.0 (0.3)	26.2	23.8 (0.5)	26.2	23.1 (0.4)
25	26.2	24.2 (0.4)	26.7	23.9 (0.4)	26.7	22.9 (0.5)
50	27.0	24.4 (0.4)	27.4	24.0 (0.4)	27.4	22.8 (0.5)
Acc 4Y	25.0 G_{rms}					

*RF is Rain-flow counting

10.5.3 Synthesis of Random Acceleration Vibration Specification from FDS_d

The FDS_d spectrum is computed both in the frequency and time domain. In the time domain, the rain-flow counting method (RF), as provided in Matlab®, is used. The number of data points of the FDS_d in the frequency domain is 2481 and in the time domain 1251. The number of data point in the original spectrum is namely 1251 too. The FDS_d is computed varying the amplification factor $Q = 10, 25, 50$ and the fatigue exponent $b = 4, 8, 12$. For $Q = 10$ and $b = 8$ the FDS_d are shown in Fig. 10.13a. To compute the equivalent random acceleration vibration specifications from the FDS_d, the spectra are divided into $N_{field} = 100$ fields. W_{Freq} is the synthesis of the random acceleration vibration specification taken from the FDS_d spectrum, and W_{RF} is the synthesis of $FDS_{d,RF}$. For $Q = 10$ and $b = 8$, both synthesised FDS are shown in Fig. 10.13b. In Fig. 10.13c both synthesised random acceleration vibration specifications are compared to the random acceleration spectrum of Accelerometer Acc-4Y.

All computed data are given in Table 10.2. The mean values and the standard deviations of the W_{RF} PSD spectra are calculated using ten samples. The figures in between the brackets are the standard deviations.

Again, it is noticed that the synthesised random acceleration vibration specifications W_{RF} (time domain) are below the original spectrum of accelerometer Acc-4Y.

10.5.4 Synthesis of Random Acceleration Vibration Specification from FDS_{pv}

The FDS_{pv} is quite similar to the FDS_d. The FDS_{pv} spectrum is computed both in the frequency and time domain. In the time domain, the rain-flow counting method (RF) was used. The number of data points of the FDS_{pv} in the frequency domain is 2481 and in the time domain 1251. The FDS_{pv} is computed varying the amplification factor $Q = 10, 25, 50$ and the fatigue exponent $b = 4, 8, 12$. For $Q = 10$ and $b = 8$,

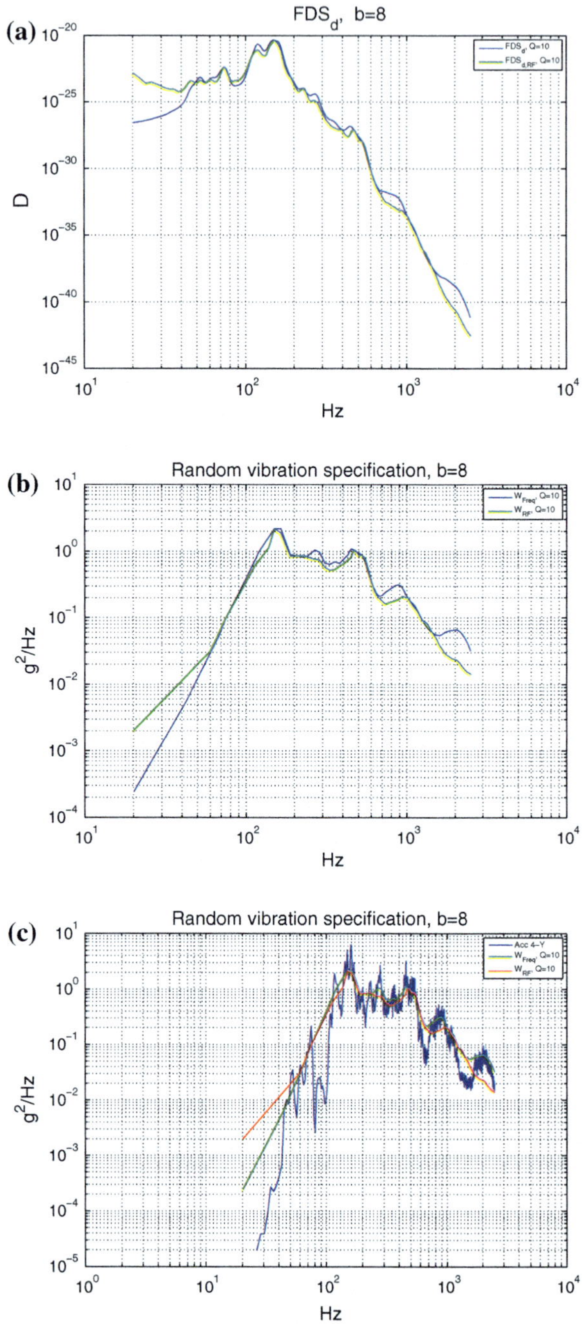

Fig. 10.13 Synthesis of random acceleration vibration specification from FDS_d (RF is rain-flow counting)

Table 10.3 Synthesis random acceleration vibration specification from FDS_{pv}

Q/r.m.s.	b					
	4		8		12	
	W_{Freq}	W_{RF} (RF*)	W_{Freq}	W_{RF}	W_{Freq}	W_{RF}
10	26.1	23.9 (0.2)	26.1	24.1 (0.4)	26.1	23.2 (0.5)
25	26.4	24.3 (0.5)	26.4	24.3 (0.4)	26.4	23.3 (0.3)
50	27.5	25.0 (0.6)	27.5	24.5 (0.4)	27.5	23.2 (0.4)
Acc-4Y	25.0 G_{rms}					

*RF is Rain-flow counting

the FDS_{pv} are shown in Fig. 10.14a. To compute the synthesised random acceleration vibration specifications, the FDS_{pv} spectra are divided in $N_{field} = 100$ fields. W_{Freq} is the synthesis of the random acceleration vibration specification taken from the FDS_{pv} spectrum, and W_{RF} is the synthesis of $FDS_{pv,RF}$. For $Q = 10$ and $b = 8$, both synthesised FDS are shown in Fig. 10.14b. In Fig. 10.14c, both synthesised random acceleration vibration specifications are compared to the random acceleration spectrum of Acc-4Y accelerometer.

All computed data are given in Table 10.3. The mean values and the standard deviations of the W_{RF} PSD spectra are calculated using ten samples. The figures in between the brackets are the standard deviations.

It is again noticed that the r.m.s. values of the synthesised random acceleration vibration specifications W_{RF} (time domain) are below the r.m.s. value of the original spectrum of accelerometer Acc-4Y.

10.5.5 Synthesis of Equivalent Random Acceleration Vibration Specification Based on Rayleigh Distribution of Peaks

The analysis of the Rayleigh distribution of peaks (damage potential) is completely done in the time domain on a transient random acceleration as illustrated in Fig. 10.10. The rain-flow method as provided by Matlab® is used to extract from the transient signal the amplitudes and corresponding number of cycles. The minimum value A_{min} of the distribution of peaks is used to calculate the number of positive crossings n_0 through A_{min}. The total time $T_{A>A_{min}} = n_o/f_i$, $i = 1, 2, \ldots$, where f_i is one the frequency in the range of the spectrum (20–2500 Hz). The number of crossings (spectrum of crossings) n_0 is shown in Fig. 10.15a. In the application, the total time is $T = 0.5004$ s, in fact the duration of the random transient signal. The total equivalent random acceleration vibration spectrum is computed from Eq. (10.33). The random acceleration vibration spectrum is divided in $N_{field} = 100$ fields, and maximum values of that spectrum in these fields and associated frequencies are indicated and applied to generate the reduced equivalent random acceleration vibration specifica-

Fig. 10.14 Synthesis of random acceleration vibration specification from FDS_{pv} (RF is rain-flow counting)

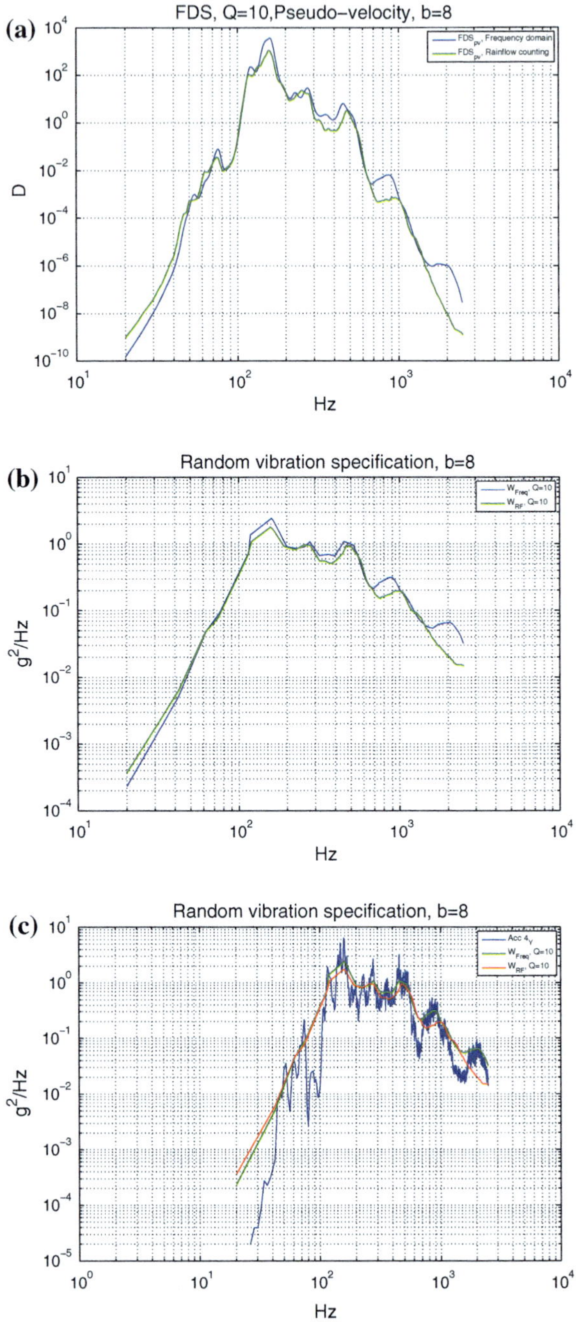

Table 10.4 Equivalent random acceleration vibration specification from Rayleigh distribution of peaks

W_{DP}	Q		
	10	25	50
Mean r.m.s. μ	26.0	25.4	24.8
Std-dev r.m.s. σ	0.7	0.4	0.2
Acc-4Y	25.0 G_{rms}		

tion W_{DP}. The full and reduced synthesised equivalent random acceleration spectra are shown in Fig. 10.15b. The comparison to the original random acceleration vibration spectrum of accelerometer Acc-4Y is provided in Fig. 10.15c. The presentation of Fig. 10.15 is based on an amplification factor $Q = 10$.

The amplification factors are varied $Q = 10, 25, 50$, and the results of the calculation are presented in Table 10.4. The mean values and standard deviations of the r.m.s values of W_{test} are calculated from ten samples.

10.5.6 All Equivalent Random Acceleration Vibration Specifications in One Plot

All synthesised equivalent random acceleration vibration specification from SRS, ESR, FDS_d, FDS_{pv}, and from the damage potential (Rayleigh distribution of peaks) are depicted in Fig. 10.16a, b. It can be seen that all equivalent random acceleration specifications have the same shape and about the same r.m.s values of the spectra; however, the r.m.s. values of the equivalent spectra computed from the frequency domain are somewhat higher than calculated in the time domain.

10.5.7 Influence Number of Fields N_{field}

The number of fields N_{field} defines to a high extent the details kept in the equivalent random acceleration vibration specification, although in combination with the number of available data points in the response spectra. In previous sections, the computations of the equivalent vibration spectra was based on a number of fields $N_{field} = 100$. The original random acceleration spectra of accelerometer Acc-4Y (Fig. 10.9) are rather smoothened in the lower-frequency range, but in the higher frequencies, the spectra still exhibit a number of peaks and valleys (i.e., Fig. 10.12). If we decrease the number of fields $N_{field} = 50, 25, 10, 5$, the equivalent random acceleration vibration specification becomes more smoothened; however, the r.m.s. value will increase, and at the other hand, the vibration specification becomes more

Fig. 10.15 Synthesis of random acceleration vibration specification from potential damage (Rayleigh distribution of peaks)

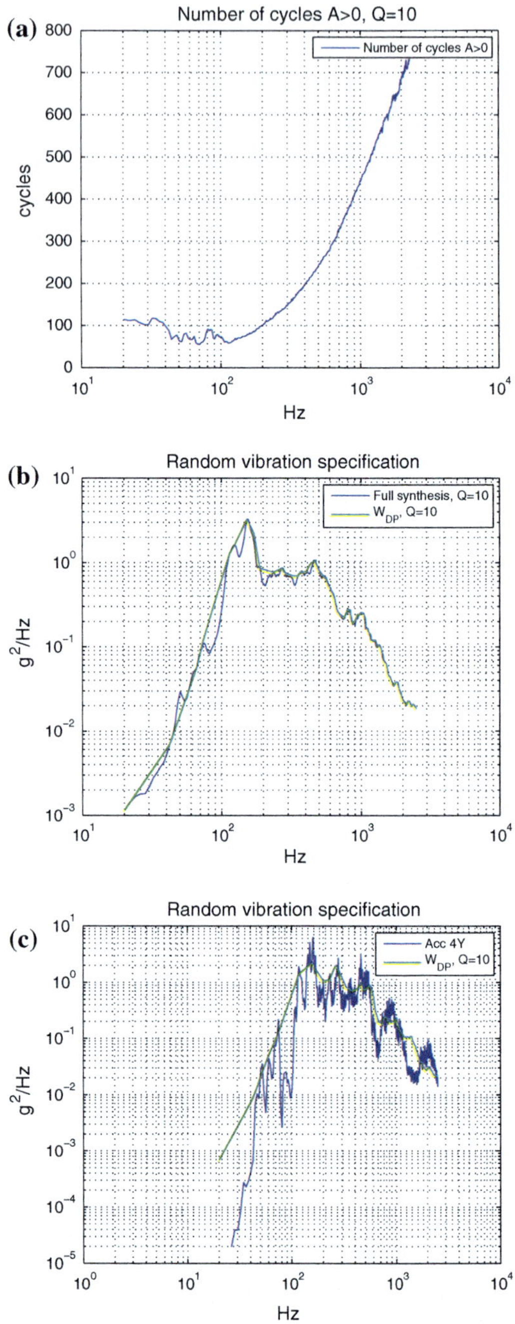

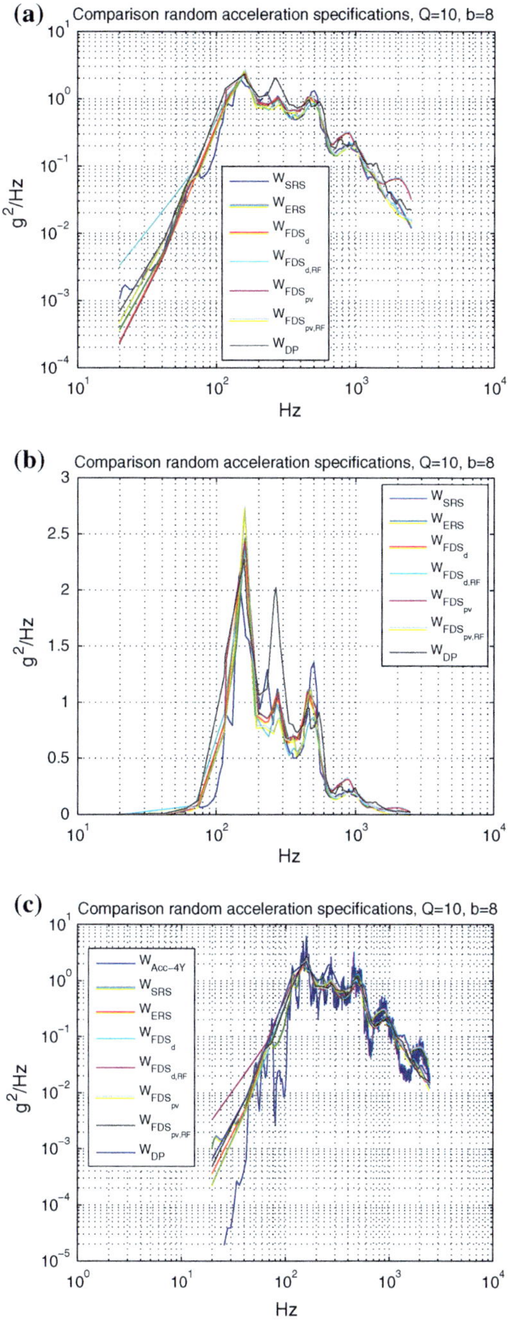

Fig. 10.16 Presentation of all equivalent random acceleration vibration specifications

Table 10.5 Equivalent random acceleration vibration spectra (r.m.s.) from ERS, SRS with varying N_{field}

	N_{field}, $Q = 10$				
	100	50	25	10	5
W_{ERS} r.m.s.	26.1	27.2	28.0	30.9	23.5
W_{SRS} r.m.s. μ	24.0	25.0	26.0	27.8	26.6
W_{SRS} r.m.s. σ	0.8	0.9	0.8	1.1	1.6

severe. This will be illustrated by an example computing the equivalent acceleration vibration specification from the original ERS assuming an amplification factor $Q = 10$. The figures are shown in Table 10.5. The mean and standard deviation of the equivalent vibration spectrum W_{SRS} are computed using ten samples. If the number of fields $N_{field} < 10$, then it becomes difficult to get an envelope of the response spectra and a r.m.s. value of the equivalent random acceleration spectrum below the r.m.s. value of the original spectrum (Fig. 10.9) will be achieved. It is recommended to keep $N_{field} = 10 - 20$ as a minimum. In Fig. 10.17a, the synthesised equivalent random acceleration vibration specifications, based on the ERS, SRS, and $N_{field} = 10$ are shown.

10.6 Chapter Summary

In this chapter, a number of damage spectra are discussed to characterise measured or computed random acceleration vibration spectra, most times defined in PSD (g^2/Hz). The characterisation is based on equivalent damage caused by extreme peaks (SRS, ERS, VRS) and Rayleigh distribution of peaks or cumulative damage (FDS), using relative displacements and pseudo-velocities. The response spectra are all based on the response of SDOF systems exited to random accelerations, both in the time and frequency domain. The principles to compute the response spectra are illustrated in the Figs. 10.2, 10.3, 10.4, and 10.5.

For random acceleration vibration, it is shown numerically that the ERS is the same as the SRS (i.e., Fig. 10.12a). The ERS is calculated in the frequency domain, but the SRS is calculated in the time domain; therefore, the computation of the ERS is more straight forward than the calculation of the SRS. Nowadays, in spacecraft structure engineering practice, the VRS is applied to compare to the SRS; however, it is recommended to use the ERS instead of the VRS, because the ERS match better with the SRS than the VRS.

The FDS_d and FDS_{pv} spectra are quite the same.

Miles' equation fulfills a key role in the synthesis process to generate equivalent random acceleration vibration specifications; however, this equation represents the r.m.s. acceleration response of a SDOF system excited at the base by white noise ran-

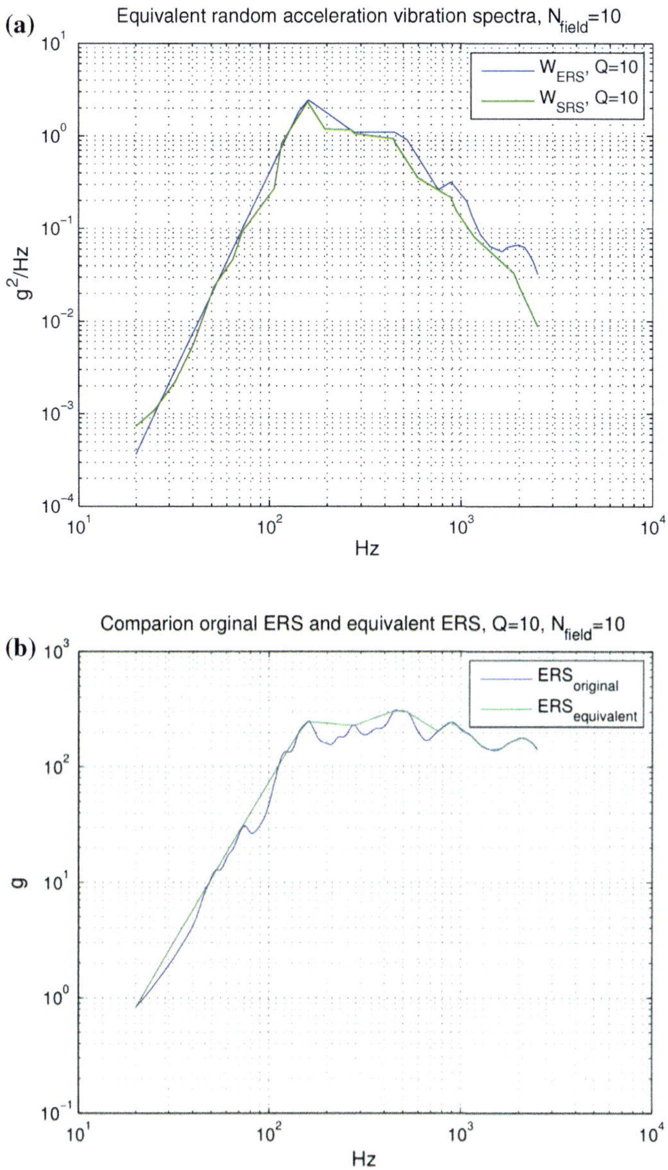

Fig. 10.17 Equivalent random acceleration vibration spectra, $N_{field} = 10$

Table 10.6 Random acceleration specification

Frequency (Hz)	g^2/Hz
20–90	3 dB/oct
90–400	0.1
400–2000	−6 dB/oct
G_{rms} (g)	?

dom acceleration, which is an idealization of the real random acceleration spectrum. The application of Miles' equation in the synthesis process is an approximation!

An envelope of the damage response spectrum is achieved by dividing the spectrum into a number of fields N_{field}. The lower N_{field} to more severe and smoother the equivalent random acceleration vibration specification becomes. This process is shown in Fig. 10.11.

All damage response spectra result in about the same equivalent random acceleration vibration specifications (see Fig. 10.16).

The methods described in this chapter are applied to a real life problem, the random acceleration spectrum (PSD) were measured during an acoustic noise test of an accelerometer placed on the outside panel of the VAATMLDS bread board solar array wing.

To handle uncertainties in damping and material properties, the amplification factor Q and the Basquin's exponent b in the s-N curve have to be varied; in general, $Q = 10, 25, 50$ and $b = 4, 8, 12$. The worst-case equivalent random acceleration vibration specification shall be selected; however, engineering judgement of the engineer is still needed.

The synthesised random transient signal derived from the Acc-4Y random spectrum has an invariant mean and standard deviation; however, the distribution of the peaks and the skewness and kurtosis show small variations.

Problems

10.1 A random acceleration specification is given in Table 10.6.

- Calculate G_{rms} of the random acceleration specification.
- Generate the SRS(f) associated with the random acceleration specification, $Q =$ 10 (see Appendices C and D).
- Generate the VRS(f) associated with the random acceleration specification, $Q =$ 10.
- Generate the ERS(f) associated with the random acceleration specification, $Q =$ 10.

Answers: $G_{rms} = 8.20\,\text{g}$,

Table 10.7 Random acceleration specification

Spectrum I		Spectrum II	
Frequency (Hz)	g^2/Hz	Frequency (Hz)	g^2/Hz
20–90	3 dB/oct	20–100	3 dB/oct
80–400	0.1	100–600	0.1
400–2000	−6 dB/oct	600–2000	−3 dB/oct
G_{rms} (g)	8.23	G_{rms} (g)	11.27

10.2 A component has a natural frequency of 90 Hz and an amplification factor of Q = 10. It has been successfully qualified to the spectrum I, Table 10.7. However, in the course of the project, the design requirements are changed. The component must now be tested to spectrum II, Table 10.7. The project management tries to avoid additional testing despite the new requirement. How could you justify that the original test is sufficient to meet the revised test specification? This is a typical problem in the "real-world" sense.[1]

- Generate the VRS(f) for both spectra I and II and plot in those in one frame.
- Calculate the PSD of spectrum II at 90 Hz.
- Calculate for the component VRS(90) values for both spectra I and II, respectively.
- What is your conclusion?

Answers: PSD = 0.09 g^2/Hz, VRS = 45.67, 33.84 g.

References

1. Girard A, Imbert JF, Moreau D (1989) Derivation of european satellite equipment test specification from vibro-acoustic test data. Acta Astronaut 10(10):797–803
2. Wijker JJ, Ellenbroek MHM, de Boer A (2013) Characterization and synthesis of random acceleration vibration specifications. In: Papadrakakis MM, Lagaros ND, Plevris V (eds) COMPDYN 2013, 4th ECCOMAS thematic conference on computational methods in structural dynamics and earthquake engineering, 12–14 June, Kos Island, Greece
3. Kelly RD, Richman G (1969) Principles and techniques of shock data analysis. SVM. The Shock and Vibration Information Center, US, MOD
4. Lalanne C (2002) Mechanical vibration and shock, specification development, vol V. HPS. ISBN 1-9039-9607-4
5. Irvine T (2009) An introduction to the vibration response spectrum, revision d. Vibrationdata, 16 June 2009
6. DiMaggio SJ, Sako BH, Rubin S (2003) Analysis of non-stationary vibroacoustic flight data using a damage-potential basis. J Spacecr Rocket 40(5):682–689
7. McNeill SI (2008) Implementing the fatigue damage spectrum and fatigue equivalent vibration testing. In: Sound and vibration, 79th shock and vibration symposium shock and vibration symposium, Orlando, Florida, 26–30 October 2008

[1]This problem is based on a course about response spectra given by www.vibrationdata.com.

8. Wirsching PH, Paez TL, Ortiz H (1995) Random vibrations, theory and practice. Wiley, New York. ISBN 0-471-58579-3
9. Miles JW (1954) On structural fatigue under random loading. J Aeronaut Sci 21(11):753–762
10. Biot MA (1933) Theory of elastic systems vibrating under transient impulse with an application to eartquake-proof buildings. Proc Natl Acad Sci 19(2):261–268
11. Biot MA (1941) A mechanical analyzer for the prediction of earthquake stresses. Bull Seismol Soc Am 31(2):150–171
12. Wijker JJ (2008) Spacecraft structures. Springer, Berlin. ISBN 978-3-540-755524
13. Smallwood DO (1981) An improved recursive formula for calculating shock response spectra. Shock Vib Bull 51:4–10
14. Ahlin K (2006) Comparison of test specifications and measured field data. Sound Vib 40:22–24
15. Tůma J, Koči P (2009) Calculation of shock response spectrum. In: Colloqium, 2–4 February 2009
16. Halfpenny A, Walton TC (2010) New techniques for vibration qualification of vibrating equipment on aircraft. Aircraft airworthiness & sustainment
17. Ochi MK (1981) Principles of extreme values statistics and their application,. In: Extreme loads response analysis symposium, pp 15–30. The societvof naval architectsand marine engineers One WorldTradeCenter, Suite1369, New York, N.Y.10048, 19–20 October 1981
18. Preumont A (1985) On the peak factor of stationary gaussian processes. J Sound Vib 100(1):15–34
19. Halfpenny A (2006) Methods for accelerating dynamic durability tests. In: 9th international conference recent advances in structural dynamics, Southampton, UK
20. Benasciutti D (2004) Fatigue analysis of random loadings. PhD thesis, University of Ferrara, Department of Engineering
21. Eaton R, Caro E (2012) Test tailoring methodology for equipment qualification. In: 2012 spacecraft and launch vehicle dynamic environments workshop, 19–21 June, El Segundo, CA, USA
22. Irvine T (2012) An alternate damage potential method for enveloping nonstationary random vibration. In: Spacecraft and launch vehicle dynamic environments workshop, p 28, El Segundo, CA, USA, 19–21 June 2012
23. Wijker JJ (2009) Random vibrations in spacecraft structures design, theory and applications. Number SMIA 165 in solid mechanics and its applications. Springer, Berlin. ISBN 978-90-481-2727-6
24. Dunham W (1999) Euler the master of us all. Number 22 in The Dolciani mathematical expositions. The Mathematical Society of America. ISBN 0-88385-328-0
25. Henderson GR, Piersol AG (2003) Evaluating vibration environments using the shock response spectrum. Sound Vib 37:18–20
26. Lalanne C (2002) Mechanical vibration nd shock, fatigue damage, vol IV. HPS. ISBN 1-9039-9606-6
27. Madayag AF (1969) Metal fatigue: theory and design. Wiley, New York. ISBN 471-56315-3
28. Crandall SH, Mark WD (1973) Random vibration in mechanical systems. Academic Press
29. Amzallag C, Geray JP, Robert J, Bahuaud JL (1994) Standardization of rain flow counting method for fatigue analysis. Fatigue, 16:287–293
30. Nieslony A (2009) Implementing the fatigue damage spectrum and fatigue equivalent vibration testing. Mech Syst Signal Process 23(8):2712–2721
31. Crandall SH (1962) Relation between strain and velocity in resonant vibration. J Acoust Soc Am 34(12):160–1961
32. Irvine T (2012) Shock and vibration stress as a function of velocity, Vibrationdata, 21 May 2012
33. Wijker JJ (2011) Vibro-acoustic analysis test methods for large deployable structures. Final presentation, ESA/TRP Study contract AO/1-5659/08/NL/EM, Noordwijk, 20 February 2011
34. IABG (2012) Iabg, space division, survey of facilities. Technical report TN-TR-1000, Issue 12, IABG, May 2012

35. Shinozuka M, Jan CM (1972) Digital simulation of random processes and its applications. J Sound Vib 25(1):111–128
36. Miles RN (1992) Effect of spectral shape on acoustic fatigue life estimation. J Sound Vib 153(2):376–386
37. Steinwolf A, White RG (1997) Probability density functions of acoustically induced strains in experiments with composite plates. AIAA 35(12):1853–1861

Chapter 11
Example Problems

Abstract A number of typical spacecraft structure-related applications of Miles' equation are worked out and explained. To gain more understanding out of the problems favorite finite element software packages may be applied to compare the answers.

Keywords Spacecraft structure-related applications

11.1 Introduction

In this chapter, a number of typical examples is worked out and explained. To gain more understanding out of the problems your favorite software packages may be applied to check the answers.

11.1 The payload adapter of a L/V is a cylinder with a radius of $r = 2$ m and a height $H = 2$ m. The cylinder is made of sandwich structure with a face sheet thickness $t = 0.5$ mm and the height of the honeycomb core $h = 10$ mm. The face sheets are made of an Al-alloy with Young's modulus $E = 70$ GPa. A S/C has a mass of $M = 300$ kg at a distance of 3.5 m from the origin and is rigidly connected to the top of the cylinder. The payload adapter is illustrated in Fig. 11.1.

Calculate the r.m.s. acceleration of COG of the S/C when the base is excited by an random acceleration $W_{\ddot{U}} = 0.004\,\text{g}^2/\text{Hz}$ in a frequency range of 20–500 Hz.

To keep the dynamic response analysis simple the cylinder is considered to be a cantilevered beam with bending stiffness EI and length $L = 2$ m. The S/C has a rigid offset $e = 1.5$ m. An unit load $F = 1$ N is applied at the location of the S/C. That means that at the tip of the beam a shear force F and bending moment Fe apply. The displacement w and rotation ϕ at the tip of the beam can be calculated with the Myosotis equations [1]

$$w = \frac{FL^3}{3EI} + \frac{FeL^2}{2EI},$$
$$\phi = \frac{FL^2}{2EI} + \frac{FeL}{EI}. \tag{11.1}$$

J. Wijker, *Miles' Equation in Random Vibrations*, Solid Mechanics and Its Applications 248, https://doi.org/10.1007/978-3-319-73114-8_11

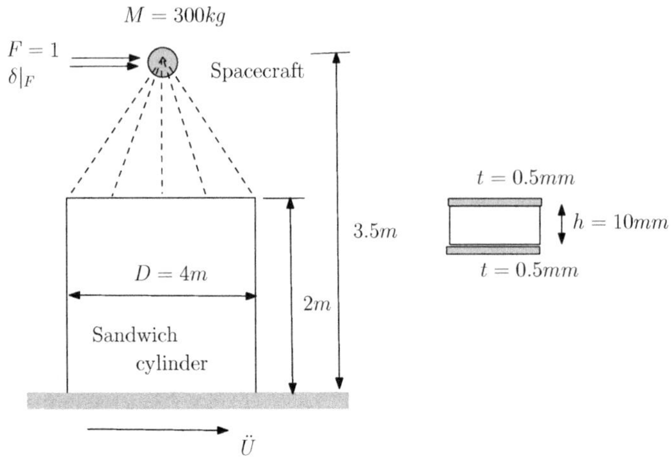

Fig. 11.1 S/C payload adapter

The total displacement δ at the location of the S/C (center of mass, COM) is given by

$$\delta = w + \phi e. \tag{11.2}$$

The stiffness at the location of the S/C with respect to fixation of the beam can be calculated with

$$k = \frac{1}{\delta|_{F=1}}. \tag{11.3}$$

When the mass of the cylinder is neglected the approximation of the natural frequency f_n is

$$f_n \approx \frac{1}{2\pi}\sqrt{k/M}\ \text{Hz}. \tag{11.4}$$

The second moment of area of the sandwich cylinder if $r = D/2$ is the outer radius can be obtained by the following expression

$$I = 2\pi t\left(\left(r - \frac{t}{2}\right)^3 + (r - h - t)^3\right)\ \text{m}^4. \tag{11.5}$$

The r.m.s. acceleration of the COM of the S/C can be calculated with Miles' equation, when the damping ratio is $\zeta = 2\%$.

$$G_{rms} = \sqrt{\frac{\pi}{2} f_n Q W_{\ddot{U}}(f_n)} = 4.8477\ \text{g}, \tag{11.6}$$

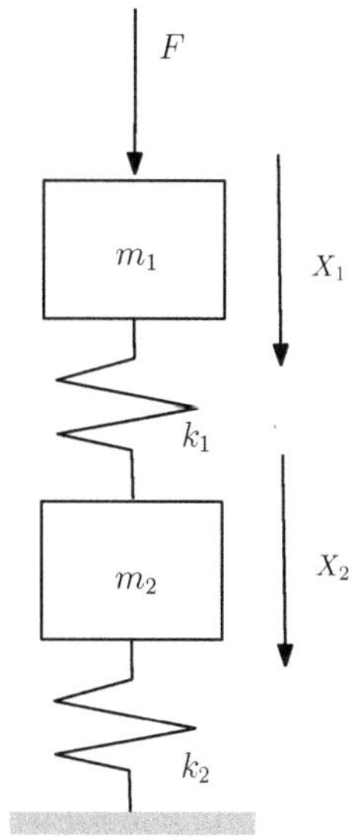

Fig. 11.2 Dynamic system

where the following numerical value are substituted:

- The second moment of area of the cylinder is $I = 0.0499$ m^4.
- The displacement of the COG of the S/C $\delta|_{F=1} = 3.7723 \times 10^{-9}$ m.
- The natural frequency $f_n = 149.6087$ Hz.
- $W_{\ddot{U}}(f_n) = 0.004$ g^2/Hz.
- Amplification factor $Q = 10$.

11.2 A dynamic system consists of two coupled SDOF elements and loaded by a random force F. This system is shown in Fig. 11.2.

The equations of motion in matrix notation are given by

$$[M](\ddot{X}) + [C](\dot{X}) + [K](X) = (F), \tag{11.7}$$

where $[M]$, $[C]$, $[K]$ are the mass, damping and stiffness matrices and $(\ddot{X})$, $(\dot{X})$, (X) the acceleration velocity and displacement vectors. (F) is the vector of external applied forces. The damping matrix $[C]$ will be constructed later on.

The discrete masses are $m_1 = 100$ kg, $m_2 = 150$ kg, and the spring stiffnesses are $k_1 = 50 \times 10^6$ N/m, $k_2 = 75 \times 10^6$ N/m. The modal damping ratio is $\zeta = 0.05$. The mass and stiffness matrices are given by:

$$[M] = \begin{bmatrix} m_1 & 0 \\ 0 & m_2 \end{bmatrix}, [K] = \begin{bmatrix} k_1 & -k_1 \\ -k_1 & k_1 + k_2 \end{bmatrix}. \tag{11.8}$$

After substituting the given values for the discrete masses and spring stiffnesses, the associated undamped natural frequencies (f_n), real vibration modes $[\Phi]$ (Hz), and

Table 11.1 PSD of applied force F

Frequency (Hz)	PSD W_F (N^2/Hz)
0–20	0
20–2000	10000
> 2000	0
F_{rms}	4450.8 N

generalized masses $[mg]$ can be computed

$$(f_n) = \begin{pmatrix} 75.6125 \\ 167.5007 \end{pmatrix}, [\Phi] = \begin{bmatrix} -0.0830 & -0.0558 \\ -0.0455 & 0.0678 \end{bmatrix},$$
$$[mg] = [\Phi]^T M[\Phi] = \begin{bmatrix} 1.0000 & 0.0000 \\ 0.0000 & 1.0000 \end{bmatrix}. \tag{11.9}$$

The orthogonal damping matrix $[C]$ can be obtained from the following expression [2]

$$[C] = \sum_{k=1}^{2} 2\zeta_k \omega_k [M](\phi_k)(\phi_k)^T [M] = 10^4 \begin{bmatrix} 0.6547 & -0.3273 \\ -0.3273 & 1.3093 \end{bmatrix} \tag{11.10}$$

where $\omega_k = 2\pi f_n$ and $\zeta_k = \zeta$ the modal damping ratio is associated with mode (ϕ_k).

The PSD W_F of the random force F is specified in Table 11.1.

The computation of the r.m.s. displacements and r.m.s. velocities of the discrete masses m_1, m_2 is done applying three different solution methods:

Time domain The PSD of the applied force is transformed into the time domain as discussed in Appendix C. The Newmark-beta method is applied to perform the response analyses in the time domain [3–5].

Frequency domain The spectral method is used to compute the responses. This approach is discussed in Appendix A.

Lyapunov equation The 'Stochastic dual of the direct method of Lyapunov' is applied to solve the mean square responses of the displacements and velocities when the MDOF system is exposed to random white noise forces [6, 7]. The mean square solution of the acceleration does not exist because a white noise force is assumed.

Time Domain Response Analysis

The PSD function of the random force $W_F = 10000$ N^2/Hz and is constant between 20–2000 Hz. Outside that frequency range $W_F = 0$. The PSD function W_F is transformed into the time domain with aid of the Fourier Transform (Appendix C) [8]. The PSD and the time series of F are shown in Fig. 11.3. The density function of the time series of F is normal distributed and is shown in Fig. 11.4.

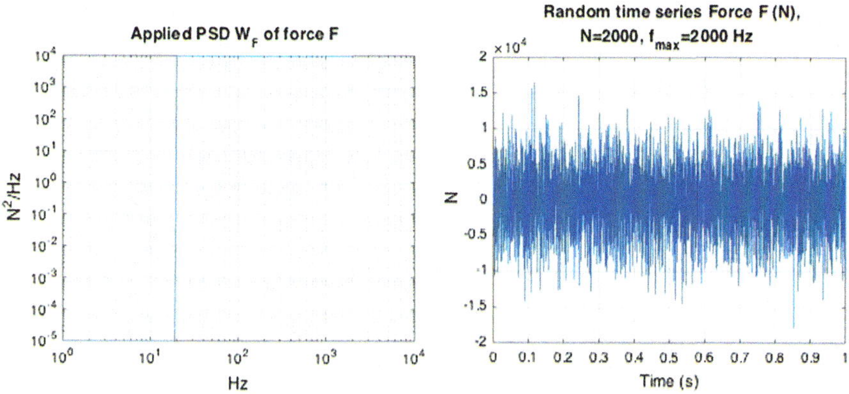

(a) PSD (b) Time domain

Fig. 11.3 Random load F in frequency and time domain

Fig. 11.4 Density function of random load F in time domain

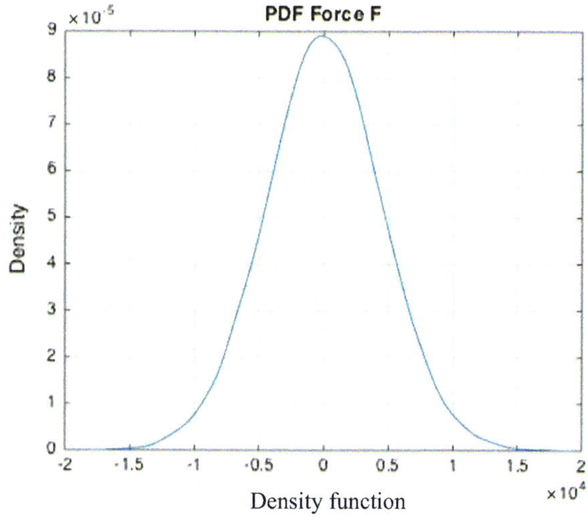

The response analysis in the time domain is done applying the well-known Newmark-beta method with $\gamma = 0.5$ and $\beta = 0.26$, [4, 5]. The results of the time domain numerical response analyses are shown in Fig. 11.5.

The time domain responses are transformed into PSD functions applying the inverse Fourier Transform as discussed in Appendix C, [8]. The PSD functions of the responses of the masses m_1, m_2 are shown in Fig. 11.6

The r.m.s. values of the displacements and velocities are presented in the first column of Table 11.2. The r.m.s. values are computed from the standard deviation of the responses.

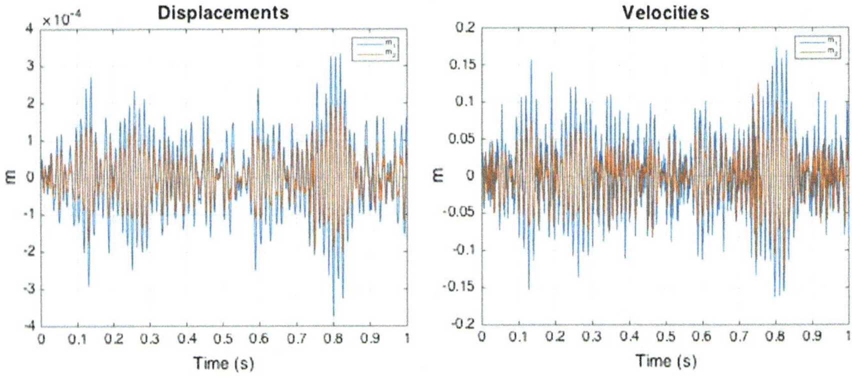

(a) Displacements (b) Velocities

Fig. 11.5 Random displacement and velocity responses in the time domain

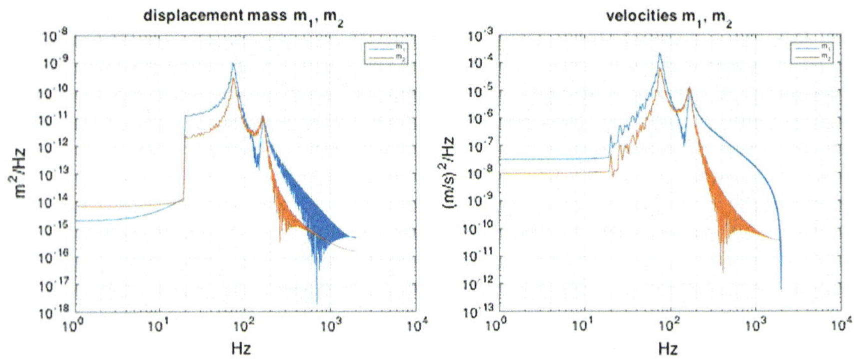

(a) PSD Displacements (b) PSD Velocities

Fig. 11.6 PSD of random displacement and velocity responses

Table 11.2 Responses of discrete masses m_1, m_2

Responses	Time domain	Frequency domain	Lyapunov equation
$x_{1,rms}$ (m)	1.0590e-04	1.0528e-04	1.0083e-04
$x_{2,rms}$ (m)	6.0068e-05	5.9765e-05	5.7324e-05
$\dot{x}_{1,rms}$ m/s	0.0524	0.0523	0.050
$\dot{x}_{2,rms}$ m/s	0.0329	0.0328	0.0316

Frequency Domain Response Analysis

The PSD of the random responses (displacement, velocity) of the masses m_1, m_2 of the dynamic system (Fig. 11.2) exposed to a random force are presented in Fig. 11.7. The method to calculate the PSD functions of the responses is traditionally done in the frequency domain [9].

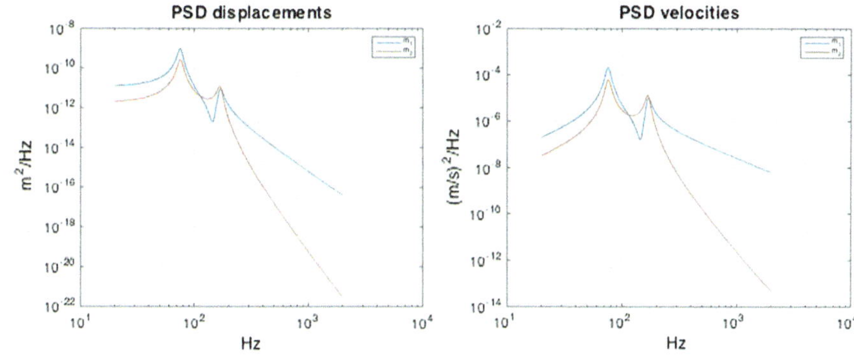

(a) PSD Displacements (b) PSD Velocities

Fig. 11.7 PSD of random displacement and velocity responses

The PSD values of the responses of masses m_1, m_2 are depicted in Fig. 11.7. The responses are computed between 20 and 2000 Hz.

The r.m.s. values of the displacement vector X and velocity vector $\dot{X}$ are presented in the second column of Table 11.2.

Lyapunov Equation Solution [10]

We transfer the matrix equation of motion Eq. (11.7) into a state space equation. The space state variables (Y_1), (Y_2) are made equal to the physical displacement vector $X = (X_1, X_2)^T$ and velocity vector $\dot{X} = (\dot{X}_1, \dot{X}_2)^T$,

$$\begin{pmatrix} Y_1 \\ Y_2 \end{pmatrix} = \begin{pmatrix} X \\ \dot{X} \end{pmatrix}. \tag{11.11}$$

Equation (11.7) can be rewritten into the state space

$$\begin{pmatrix} \dot{Y}_1 \\ \dot{Y}_2 \end{pmatrix} = \begin{pmatrix} [0] & [I] \\ -[M]^{-1}[C] & -[M]^{-1}[K] \end{pmatrix} \begin{pmatrix} Y_1 \\ Y_2 \end{pmatrix} = \begin{pmatrix} [M]^{-1}(F) \\ 0 \end{pmatrix}, \tag{11.12}$$

or

$$\begin{pmatrix} \dot{Y}_1 \\ \dot{Y}_2 \end{pmatrix} = [A] \begin{pmatrix} Y_1 \\ Y_2 \end{pmatrix} = (B)(F). \tag{11.13}$$

The following Lyapunov equation is now to be solved:

$$[A]^T[P] + [P][A] = \frac{W_F}{2}(B)(B)^T, \tag{11.14}$$

where

$$[P] = \begin{bmatrix} < X_1X_1 > & < X_1X_2 > & < X_1\dot{X}_1 > & < X_1\dot{X}_2 > \\ < X_2X_1 > & < X_2X_2 > & < X_2\dot{X}_1 > & < X_2\dot{X}_2 > \\ < \dot{X}_1X_1 > & < \dot{X}_1X_2 > & < \dot{X}_1\dot{X}_1 > & < \dot{X}_1\dot{X}_2 > \\ < \dot{X}_2X_1 > & < \dot{X}_2X_2 > & < \dot{X}_2\dot{X}_1 > & < \dot{X}_2\dot{X}_2 > \end{bmatrix} \tag{11.15}$$

After the numerical evaluation of the Lyapunov equation with $W_F = 10000\,\mathrm{N^2/Hz}$ the correlation matrix $[P]$ is

$$[P] = \begin{bmatrix} 0.000000011316 & 0.000000005805 & 0.000000000000 & 0.000000211118 \\ 0.000000005805 & 0.000000003609 & -0.000000211118 & 0.000000000000 \\ 0.000000000000 & -0.000000211118 & 0.002748455095 & 0.001084178962 \\ 0.000000211118 & 0.000000000000 & 0.001084178962 & 0.001077268472 \end{bmatrix}$$

The r.m.s. values of the diagonal terms are presented in the third column of Table 11.2.

Summary

In this section, the results of the response analyses are summarized in Table 11.2.

The r.m.s. responses calculated using the Lyapunov equation are some what lower.

11.3 A cantilevered beam is loaded by an uniform random distributed load $q(t)$ is shown in Fig. 11.8. The bending stiffness of the beam is EI, and the mass per unit of length is m and the total length is L.

Calculate the random response $W(x,t)$ using Miles' equation. The displacement $W(x,t)$ is divided into a function $\phi(x)$, only dependent upon the geometry and a time-dependent generalized coordinate $\eta(t)$, such that

$$W(x,t) = \eta(t)\phi(x), \tag{11.16}$$

where the assumed mode $\phi(x)$ is given by [11]

$$\phi(x) = x^2(6L^2 - 4xL + x^2). \tag{11.17}$$

The undamped equation of motion of the vibrating loaded beam is expressed as

$$EI\frac{\partial^4 W(x,t)}{\partial x^4} + m\frac{\partial^2 W(x,t)}{\partial t^2} = q(t). \tag{11.18}$$

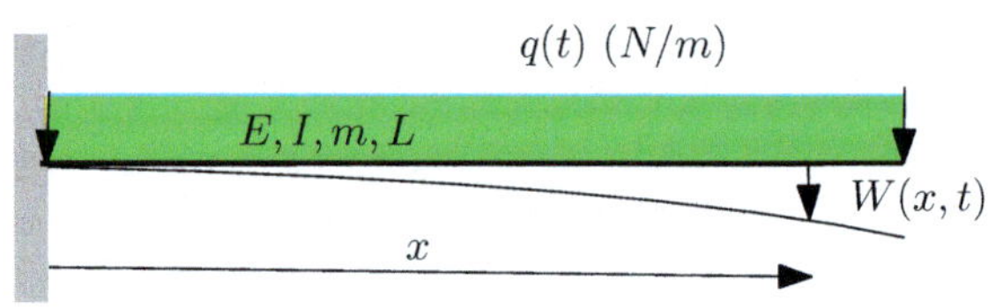

Fig. 11.8 Cantilevered beam load by uniform random load

When we substitute $W(x,t) = \eta(t)\phi(x)$ into the equation of motion of the beam and integrating over the length of the beam we obtain

$$EI\eta(t)\int_0^L \phi(x)\phi^{IV}(x)dx + m\ddot{\eta}(t)\int_0^L \phi^2(x)dx = q(t)\int_0^L \phi(x)dx, \quad (11.19)$$

or further worked out

$$\frac{162EI}{13mL^4}\eta(t) + \ddot{\eta}(t) = q(t)\frac{27}{52mL^4}. \quad (11.20)$$

Adding ad hoc the damping term, we finally get the damped equation of motion of a SDOF system

$$\ddot{\eta}(t) + 2\zeta\omega_n\dot{\eta}(t) + \omega_n^2\eta(t) = \Gamma q(t). \quad (11.21)$$

Using Miles' equation, we calculate the r.m.s. value of the generalized coordinate $\eta(t)$

$$\eta_{rms} = \frac{\Gamma}{(2\pi f_n)^2}\sqrt{\frac{\pi}{2}Qf_n W_q(f_n)}. \quad (11.22)$$

The r.m.s. value of the physical displacement $W(x,t)$ can now obtained by the following expression

$$w(x)_{rms} = \eta_{rms}\phi(x). \quad (11.23)$$

The r.m.s value of the bending moment at the root of the beam $M(0)_{rms}$ is given by

$$M(0)_{rms} = EI\frac{d^2w_{rms}}{dx^2}|_{x=0} = 12EIL^2\eta_{rms}, \quad (11.24)$$

and the r.m.s value of the shear force at the root of the beam $D(0)_{rms}$ is given by

$$D(0)_{rms} = EI\frac{d^3w_{rms}}{dx^3}|_{x=0} = 24EIL\eta_{rms}. \quad (11.25)$$

11.4 An instrument is modeled by a simple dynamic system consisting of two coupled SDOF elements and is excited at the base by an enforced acceleration $\ddot{U}$. This system is shown in Fig. 11.9.

The discrete masses m_1, m_2 are both 10 kg. The total mass of the system is $M_{tot} = 20$ kg. The spring stiffness of the spring $k_1 = 12 \times 10^6$ N/m and the spring stiffness of spring $k_2 = 1.5k_1$. The modal damping ratio $\zeta = 0.05$. The instrument is statically designed against a static load factor specified in the mass acceleration curve (MAC), which is discussed in detail in [13]. In our case, the static design load factor $\ddot{y}_{stat} = 30$ g. The MAC is shown in Fig. 11.10.

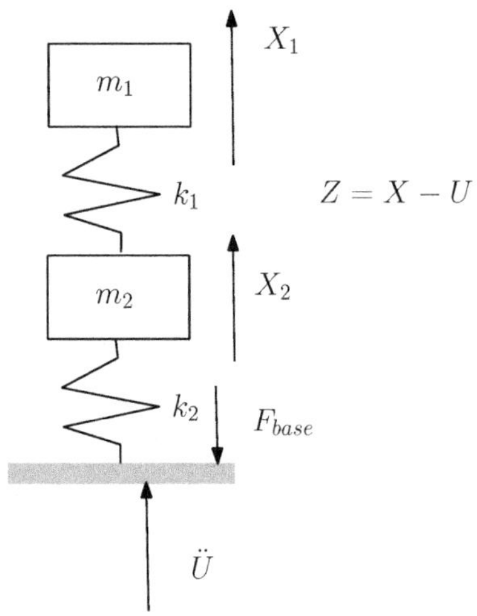

Fig. 11.9 Two SDOF systems

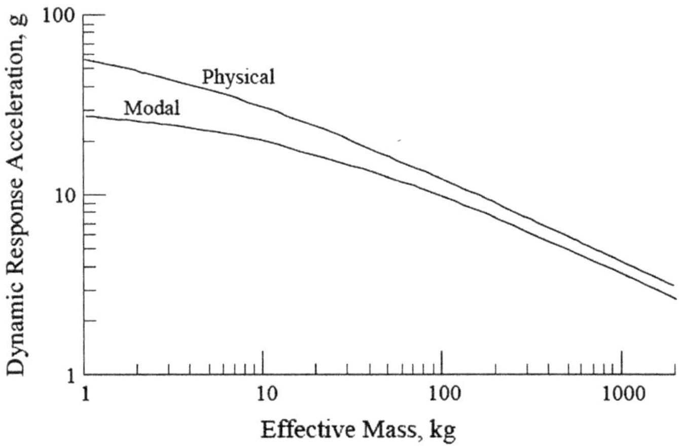

Fig. 11.10 Mass acceleration curve [12]

The equations of motion in matrix notation are given by

$$[M](\ddot{Z}) + [C](\dot{Z}) + [K](Z) = -[M](RB)\ddot{U}, \tag{11.26}$$

where $[M]$, $[C]$, $[K]$ are the mass, damping and stiffness matrices and $(\ddot{Z})$, $(\dot{Z})$, $(Z) = (X) - (RB)U$ the relative acceleration velocity and displacement vectors. $\ddot{U}$ is the enforced acceleration and $(RB) = (1, 1)^T$ is the rigid body mode. The orthogonal damping matrix $[C]$ can be obtained from Eq. (11.10)

$$[C] = \sum_{k=1}^{2} 2\zeta_k \omega_k [M](\phi_k)(\phi_k)^T [M] = 10^3 \begin{bmatrix} 0.9992 & -0.4491 \\ -0.4491 & 1.6728 \end{bmatrix}. \tag{11.27}$$

After substituting the given values for the discrete masses and spring stiffnesses in Eq. (11.8), the undamped natural frequencies (f_n) (Hz), real vibration modes $[\Phi]$, and generalized masses $[mg]$ can be computed

$$\begin{aligned} (f_n) &= \begin{pmatrix} 123.28 \\ 301.99 \end{pmatrix}, [\Phi] = \begin{bmatrix} -0.2828 & -0.1414 \\ -0.1414 & 0.2828 \end{bmatrix}, \\ [mg] &= [\Phi]^T M[\Phi] = \begin{bmatrix} 1.0000 & 0.0000 \\ 0.0000 & 1.0000 \end{bmatrix}. \end{aligned} \tag{11.28}$$

The modal effective masses M_{eff} associated with the first and second vibration modes are

$$(M_{eff}) = \begin{pmatrix} 18.0000 \\ 2.0000 \end{pmatrix} \text{ kg}. \tag{11.29}$$

The modal effectives mass corresponding to the first mode (ϕ_1) is dominant compared to the second mode (ϕ_2).

The PSD of the enforced acceleration $\ddot{U}$ is provided in Table 11.3.

The definition of the random load factor is the 3σ value of reaction force F_{base} divided by the total mass of the dynamical system M_{tot}.

$$\ddot{\gamma} = \frac{3F_{base,rms}}{M_{tot}}. \tag{11.30}$$

The r.m.s. value of the reaction force F_{base} will calculated by two different approaches:

The spectral method The 3σ value of the random force in spring k_2 is calculated and divided by the total mass M_{tot} kg to obtain the random load factor $\ddot{\gamma}$. The number of frequency steps is 5000, $\Delta f = 0.391$ Hz.

Miles' equation The random load factor $\ddot{\gamma}$ is calculated using Eq. (4.8).

$$\ddot{\gamma} = \frac{3}{M_{tot}} \sqrt{\sum_{k=1}^{n} \frac{\pi}{2} M_{eff}^2(f_{n,k}) f_k Q_k W_{\ddot{U}}(f_{n,k})}.$$

Table 11.3 Acceleration power spectrum $\ddot{U}$

Frequency range (Hz)	PSD (ASD) levels (g^2/Hz)
20–100	3 dB/oct.
100–300	0.1
300–2000	−6 dB/oct.
G_{rms}	7.09

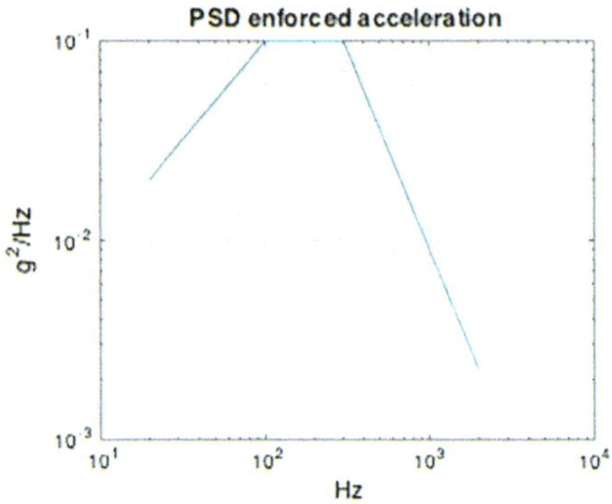

(a) PSD Enforced accelerations

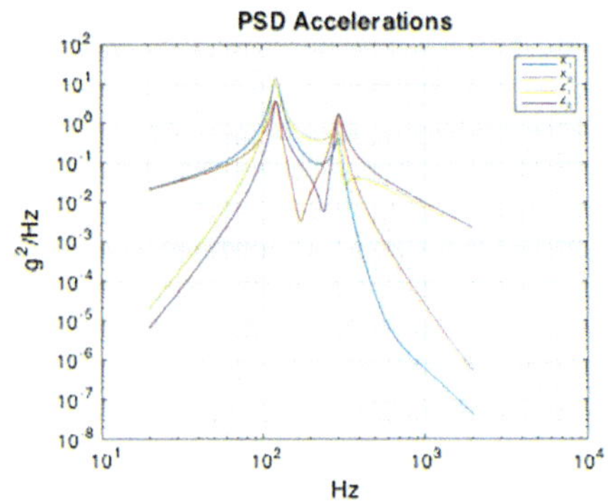

(b) PSD Accelerations

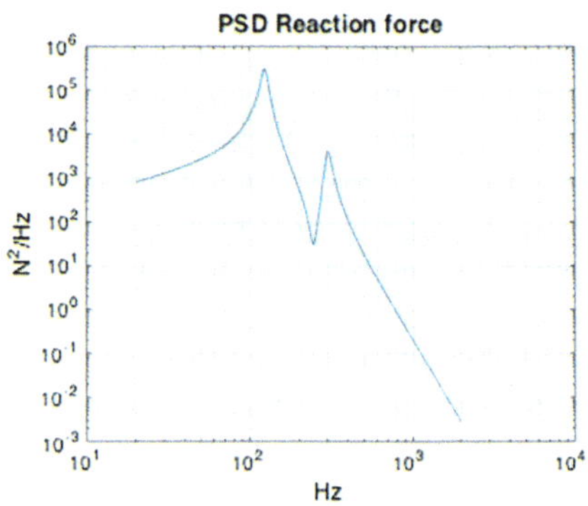

(c) PSD Reaction force

Fig. 11.11 PSD enforced accelerations, acceleration responses and reaction force in spring k_2

Spectral method

The equations to be solved in the frequency domain are:

$$
\begin{aligned}
&(-\omega^2[M] + j\omega[C] + [K])(Z(\omega)) = -[M](RB),\\
&(\ddot{X}(\omega)) = -[M]^{-1}(j\omega[C] + [K])(Z(\omega))\\
&F_{base}(\omega) = k_2 Z(2).
\end{aligned}
\tag{11.31}
$$

The PSD of the acceleration and the reaction force can be calculated with

$$
\begin{aligned}
&W_x(f) = |X(\omega)|^2 W_{\ddot{U}}(f),\ W_z(f) = |Z(\omega)|^2 W_{\ddot{U}}(f),\\
&W_{F_{base}}(f) = |F_{base}(\omega)|^2 W_{\ddot{U}}(f),
\end{aligned}
\tag{11.32}
$$

where ω is radial frequency (rad/s) and f the excitation frequency in Hz (cycles/s).

The PSD of the enforced acceleration $W_{\ddot{U}}(f)$ and the PSD of the absolute and relative accelerations, $(\ddot{X})$ and $(\ddot{Z})$ respectively, are presented in Fig. 11.11.

Table 11.4 r.m.s. values of responses

$\ddot{x}_{1,rms}$	$\ddot{x}_{2,rms}$	$\ddot{z}_{1,rms}$	$\ddot{z}_{2,rms}$	$F_{base,rms}$
(g)	(g)	(g)	(g)	(N)
17.0226	11.7348	18.7719	12.9152	5.9491×10^6

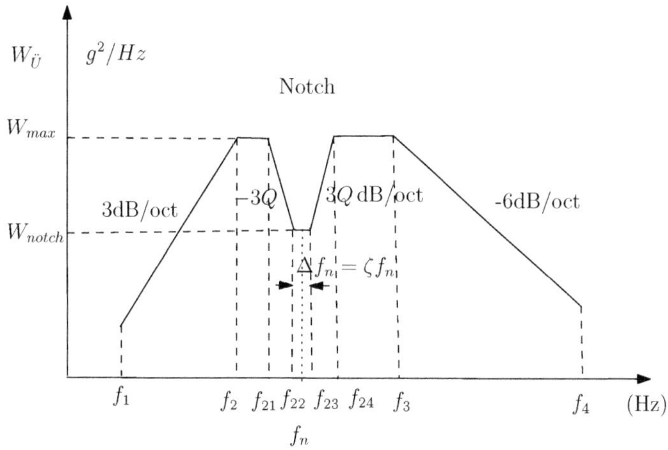

Fig. 11.12 Random enforced acceleration $W_{\ddot{U}}$ with notch

The r.m.s. values of the relative and absolute accelerations and reaction force are presented in Table 11.4.

The random load factor $\ddot{\gamma}$ can be calculated

$$\ddot{\gamma} = \frac{3F_{base,rms}}{9.81M_{tot}} = 37.2946\,\text{g}. \tag{11.33}$$

The calculated random load factor must be reduced to be compliant with the static load factor $\ddot{\gamma}_{stat} = 30\,\text{g}$. The PSD of the enforced acceleration $W_{\ddot{U}}$ must be adapted and a notch shall be defined (Fig. 11.12).

At first, we define the natural frequency and mode with the dominant modal effective mass ($M_{eff} = 18\,\text{kg}$) and the associated bandwidth $\Delta f_n = \zeta f_n$ ($\zeta = 0.05$). The slopes in the notch are $\mp 3Q$ and the frequencies shaping the notch can be obtained by the following expressions when $W_{max}/W_{notch} = 2.5$ (to be varied):

$$f_{22} = f_n\left(1 - \frac{\zeta}{2}\right),\ f_{23} = f_n\left(1 + \frac{\zeta}{2}\right)$$
$$f_{21} = f_{22}\left(\frac{W_{max}}{W_{min}}\right)^{-\frac{3}{3Q}},\ f_{24} = f_{23}\left(\frac{W_{max}}{W_{min}}\right)^{\frac{3}{3Q}}. \tag{11.34}$$

Table 11.5 Key frequencies of the notch (Hz)

	f_n	Δf_n	f_1	f_2	f_{21}	f_{22}	f_{23}	f_{24}	f_3	f_4
f (Hz)	123.28	6.16	20	100	109.67	120.20	126.36	138.49	300	2000
$W_{\ddot{U}}(f)\text{g}^2/\text{Hz}$	0.04			0.1	0.1	0.04	0.04	0.1	0.1	
G_{rms} (g)	7.01 (original 7.09)									

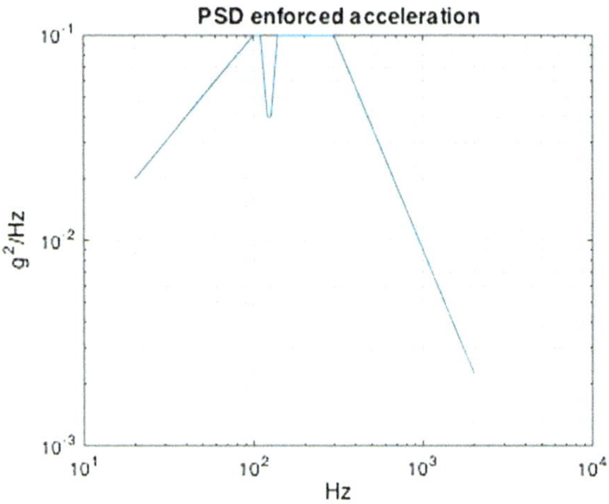

Fig. 11.13 Random enforced acceleration spectrum $W_{\ddot{U}}$ with notch

Table 11.6 r.m.s. values of responses

$\ddot{x}_{1,rms}$	$\ddot{x}_{2,rms}$	$\ddot{z}_{1,rms}$	$\ddot{z}_{2,rms}$	$F_{base,rms}$
(g)	(g)	(g)	(g)	(N)
13.6702	10.5464	15.7968	11.8809	3.7329×10^6

All numerical values of the frequencies shaping the complete spectrum of the enforced acceleration are presented in Table 11.5.

The shape of the PSD of the enforced acceleration $W_{\ddot{U}}$ is shown in Fig. 11.13.

The r.m.s. values of the relative and absolute accelerations and reaction force are presented in Table 11.6.

The random load factor $\ddot{\gamma}$ can be calculated with

$$\ddot{\gamma} = \frac{3F_{base,rms}}{9.81M_{tot}} = 29.5425\ \text{g}. \tag{11.35}$$

Table 11.7 PSD values enforced acceleration at natural frequencies

	$f_1 = 123.28$ Hz	$f_2 = 301.97$ Hz	$\ddot{\gamma}$	Remark
$W_{\ddot{U}}(f_k)$ (g^2/Hz)	0.1000	0.0987	38.1293	
$W^*_{\ddot{U}}(f_k)$ (g^2/Hz)	0.0503	0.0987	27.4144	First iteration
$W^*_{\ddot{U}}(f_k)$ (g^2/Hz)	0.0608	0.0987	30.0000	Second iteration
$M_{eff}(f_k)$ (kg)	18	2		

The random load factor is now close to the static load case of $\ddot{\gamma}_{stat} = 30\,\text{g}$, which was taken from the MAC from Fig. 11.10. Applying the adapted spectrum of the enforced accelerations over-testing of the dynamic system is prevented.

Miles' equation

$$\ddot{\gamma} = \frac{3}{M_{tot}} \sqrt{\sum_{k=1}^{n} \frac{\pi}{2} M_{eff}^2(f_{n,k}) f_{n,k} Q_k W_{\ddot{U}}(f_{n,k})} = 38.1293\,\text{g}. \tag{11.36}$$

The random load factor obtained by Miles' equations is about the same as calculated by the spectral method.

Equations (5.10) and (5.13) are used to calculate $\bar{A}$ and $W^*_{\ddot{u}}(f_k)$.

$$\bar{A} = \frac{(M_{tot}\ddot{\gamma}_{stat})^2}{\frac{9\pi}{2}\sum_{k=1}^{2}\left[M_{eff}(f_{n,k}) f_k W_{\ddot{u}}(f_{n,k})\right]} = 90.4539. \tag{11.37}$$

$$W^*_{\ddot{u}}(f_{n,k}) = \bar{A}\frac{W_{\ddot{u}}(f_{n,k})}{M_{eff}(f_{n,k})Q_k},\ k = 1, 2\ (\text{g}^2/\text{Hz}). \tag{11.38}$$

The calculation of the notch levels is performed in two iterations. The results of these two iterations are shown in Table 11.7. We notice that the PSD level of $W_{U^{\cdot\cdot}}(301.9753) = 0.0987$ did not change. The final calculation of $W^*_{U^{\cdot\cdot}}(123.2809)$ g^2/Hz is obtained by the following expression

$$W^*_{U^{\cdot\cdot}}(f_{n,1}) = \frac{\frac{(M_{tot}\ddot{\gamma}_{stat})^2}{\frac{9\pi Q}{2}} - f_{n,2} W_{U^{\cdot\cdot}}(f_{n,2}) M_{eff}^2(f_{n,2})}{f_{n,1} M_{eff}^2(f_{n,1})}\ (\text{g}^2/\text{Hz}). \tag{11.39}$$

11.5 This example is taken from ANSYS Advantage, Volume II, Issue 3, 2008 and [14]. Consider a cantilevered Al-alloy beam carrying at the tip a mass $M = 0.2\,\text{kg}$. The properties of the Al-alloy are: Young's modulus is $E = 68.9\,\text{GPa}$ and the density $\rho = 2700\,\text{kg/m}^3$. The modal damping ratio is $\zeta = 0.05$. The second moment of area is $I = bh^3/12\,(\text{m}^4)$ and the moment of resistance $W_b = 2I/h\,(\text{m}^3)$. The mass per unit length of the beam is $m = \rho bh = 02835$ (kg/m). The dimensions in Fig. 11.14 are in mm.

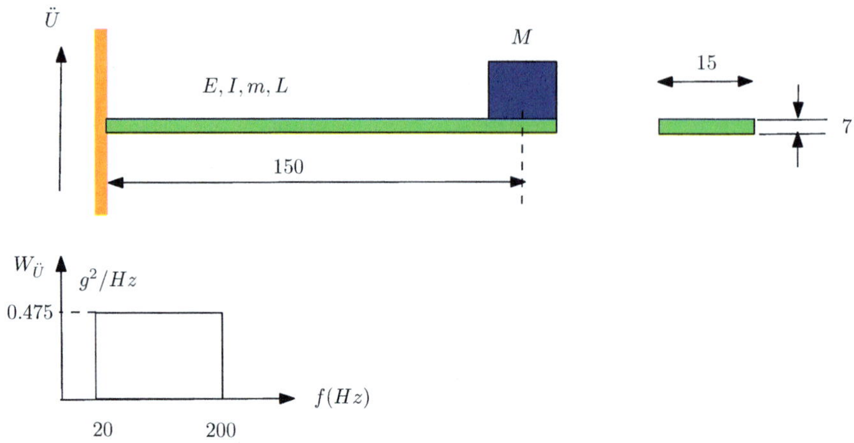

Fig. 11.14 Cantilevered beam wit mass at the tip undergoing random enforced acceleration

The three band technique [14] will be used to calculate Miner's cumulative damage ratio. The fatigue life of the shown dynamic system will be investigated when the estimated duration of all tests and launch is $T = 4 * 500\,\text{s}$, including a usual safety factor of 4.

The fundamental natural frequency f_n of the dynamic system can be calculated with the following expression [11].

$$f_n = \frac{1}{2\pi}\sqrt{\frac{1296EI}{465ML^3 + 104\,mL^4}} = 54.3085\ \text{Hz} \tag{11.40}$$

Total expected number of cycles is $\nu_0 = Tf_n$.

The G_{rms} or random load factor is now calculated with the aid of Miles' equation

$$G_{rms} = \sqrt{\frac{\pi}{2} f_n Q W_{\ddot{U}}(f_n)} = 20.1298\,\text{g}. \tag{11.41}$$

The r.m.s. bending moment at the root of the cantilevered beam is

$$M_b = \left(\frac{mL^2}{2} + ML\right) G_{rms} = 6.5540\,\text{Nm}, \tag{11.42}$$

and the corresponding r.m.s. bending stress σ_b is

$$\sigma_b = \frac{M_b}{W_b} = 53.502\,\text{MPa}. \tag{11.43}$$

The approximate number of stress cycles N, required to produce a fatigue failure in the beam for $1\sigma, 2\sigma, 3\sigma$ values of the stress can be obtained from the following provided $s - N$ curve

$$N = 1000\left(\frac{s_2}{s_1}\right)^{6.4}, \tag{11.44}$$

where

- N is number allowable stress cycles.
- $s_2 = 310\,\mathrm{MPa}$ is the stress to fail at 1000 cycles.
- $s_1 = 1\sigma_b, 2\sigma_b, 3\sigma_b$ MPa (σ_b is r.m.s. bending stress).

Miner's cumulative fatigue damage ratio is based on the idea that every stress cycle uses up a part of the fatigue life of a structure. Miner's damage cycle ratio D will be calculated with aid of Steinburg's three band technique [14]:

$$\begin{aligned} D = \sum_{k=1}^{3} \frac{n_k}{N_k} &= \nu_0\left(\frac{0.683}{N_{1\sigma_b}} + \frac{0.271}{N_{2\sigma_b}} + \frac{0.0433}{N_{3\sigma_b}}\right) \\ &= 0.0010 + 0.0325 + 0.0696 = 0.1031, \end{aligned} \tag{11.45}$$

where $\nu_0 = f_n T = 1.0862 \times 10^5$ cycles. All variables to calculate D are given in Table 11.8.

An examination of the above fatigue cycle ratio D shows that the 1σ level does very little damage even though it is in effect about 68.3% of the time. Most of the damage is generated by the 3σ level, even though it acts only about 4.33% of the time. 10% of the fatigue had been eaten up, 90% is still left.

The natural frequency of the dynamic system illustrated in Fig. 11.14 will be calculated using the finite element analysis (FEA) method, The beam will be modeled with one Bernoulli beam element with the following mass and stiffness matrices [15].

$$[M] = \frac{mL}{420}\begin{bmatrix} 156 & 22L & 54 & -13L \\ 22L & 4L^2 & 13L & -3L^2 \\ 54 & 13L & 156 & -22L \\ -13L & -3L^2 & -22L & 4L^2 \end{bmatrix} \quad [K] = \frac{EI}{L}\begin{bmatrix} 12 & 6L & -12 & 6L \\ 6L & 4L^2 & -6L & 2L^2 \\ -12 & -6L & 12 & -6L \\ 6L & 2L^2 & -6L & 4L^2 \end{bmatrix} \tag{11.46}$$

Table 11.8 Variables used in Miner's rule

s_1 (MPa)	$1\sigma_b$	$2\sigma_b$	$3\sigma_b$
N cycles ($\times 10^7$)	7.6407	0.0905	0.0068
Percentage of occurrence p (%)	68.3	27.1	4.33
$n = \nu_0 p$ cycles ($\times 10^4$)	7.4185	2.9435	0.4703

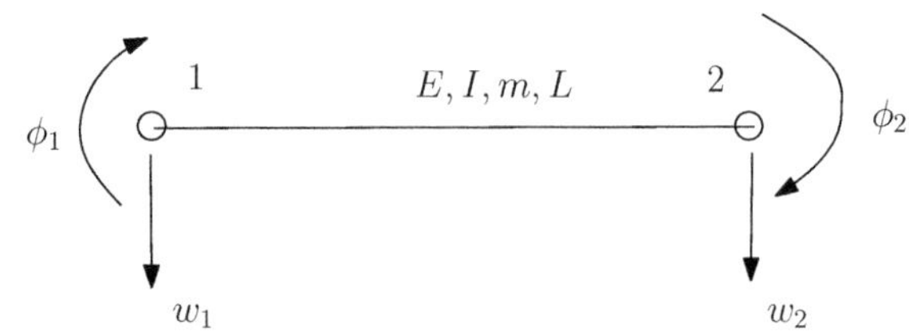

Fig. 11.15 Finite element model beam

The finite element model (FEM) is illustrated in Fig. 11.15.

To introduce the fixation, the large mass approach will be applied (mass $M_l = 10^9$ kg and second moment of mass $I_l = 10^9$ kgm^2) in the direction of w_1 and ϕ_1. The updated mass matrix is

$$[M] = \frac{mL}{420}\begin{bmatrix} 156 + M_l\frac{420}{mL} & 22L & 54 & -13L \\ 22L & 4L^2 + I_l\frac{420}{mL} & 13L & -3L^2 \\ 54 & 13L & 156 + M\frac{420}{mL} & -22L \\ -13L & -3L^2 & -22L & 4L^2 \end{bmatrix}, \tag{11.47}$$

where the mass of the box M is introduced too. To calculate the the unitary frequency response functions $H(f)$ of the accelerations at node 1, degrees of freedom (DOF) w_1 and w_2 a constant enforced acceleration $\ddot{U} = M_l$ will be applied to node 1, DOF w_1. That means that the resulting enforced acceleration is equal to one. This is the principle the large mass approach. Later on the PSD of the accelerations will be computed.

The natural frequencies f_n and corresponding vibration $[\Phi]$ modes are given hereafter

$$(f_n) = \begin{pmatrix} 0.0000 \\ 0.0000 \\ 56.2745 \\ 1503.18 \end{pmatrix}, [\Phi] = \begin{bmatrix} 0.0000 & 0.0000 & 0.0000 & 0.0000 \\ 0.0000 & 0.0000 & 0.0000 & 0.0000 \\ 0.0000 & 0.0000 & 2.1820 & 0.3899 \\ 0.0000 & 0.0000 & 21.7353 & 340.4000 \end{bmatrix} \tag{11.48}$$

The value of the lowest natural frequency obtained with more detailed FEM consisting of 10 beam elements is $f_n = 56.2739$ Hz.

The orthogonal damping matrix $[C]$ can be obtained using Eq. 11.10, with $\zeta = 0.05$.

The PSD of the accelerations of DOFs w_1 and w_2 as well as the bending moment in node 1, ϕ_1 direction are shown Fig. 11.16.

The r.m.s. acceleration of the discrete mass M is $w_{2,rms} = 20.9057$ g and the r.m.s. of the bending moment in node 1 is $M_b = 5.8107$ Nm and the r.m.s. of the bending stress is $\sigma_b = 47.4341$ MPa.

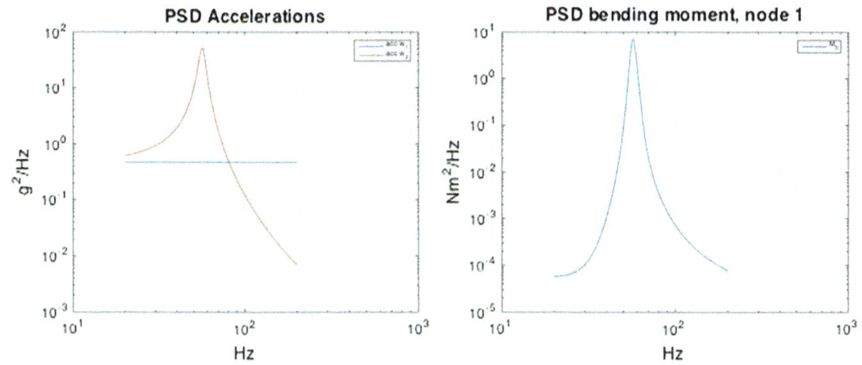

(a) Accelerations (b) Bending moment

Fig. 11.16 PSD accelerations and bending moment

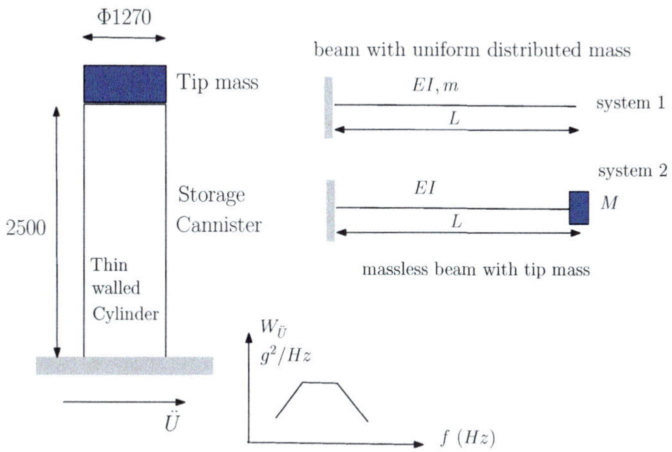

Fig. 11.17 Storage canister

The results of the computations with Miles' equation and with the simple FEM are comparable, however, as expected the bending stress σ_b calculated with the FEA is somewhat lower [10].

11.6 This problem is more or less taken from lecture notes 'Principles of Space Systems design,' University of Maryland. A storage canister for the ISS solar array deployment was launched by the Space Transportation System (STS) in a cantilevered launch configuration. The canister system is shown in Fig. 11.17. The canister is a thin-walled Al-alloy cylinder with radius $R = 635$ mm, the wall thickness is $t = 3$ mm and the length is $L = 2500$ mm. Young's modulus is $E = 70$ GPa and the density is $\rho = 2700\,\text{kg/m}^3$. The mass per unit of length $m = A\rho$ (kg/m), where A (m^2) is the area of the cross section. The tip mass is $M = 90$ kg. No stored mass is considered.

Table 11.9 PSD enforced acceleration and damping ratio

(Hz)	g^2/Hz	(Hz)	ζ
20–100	3 dB/oct.	≤ 150	0.045
100–400	0.1	150–300	0.020
400–2000	−6 dB/oct.	>300	0.005
G_{rms}	8.1733 g		

Derive the lowest natural frequency f_n (Hz) of the cantilevered canister using Dunkerley's method [16, 17]. In Fig. 11.17, two cantilevered systems are drawn:

1. System 1: cantilevered beam without the tip mass. Lowest natural frequency $\omega_{11} = 3.53624\sqrt{EI/mL^4}$
2. System 2: cantilevered massless beam with tip mass. Lowest natural frequency $\omega_{21} = \sqrt{3EI/ML^3}$

In accordance with Dunkerley's method, the lowest natural frequency of the cantilevered beam with tip mass can be approximated with the following expression

$$\frac{1}{\omega_1^2} \approx \frac{1}{\omega_{11}^2} + \frac{1}{\omega_{21}^2},$$
$$f_n = \frac{\omega_1}{2\pi} = 0.27567\sqrt{\frac{EI}{L^3(M + 0.2427mL)}} = 80.0668 \text{ Hz}. \tag{11.49}$$

The PSD of the random enforced acceleration $W_{\ddot{U}}(f)$ and the frequency dependent damping ratio ζ are provided in Table 11.9.

The PSD $W_{\ddot{U}}(f_n) = 0.0801\,g^2$/Hz and the amplification factor at f_n is $Q = 11.1111$. The $3\sigma_a$ value of the acceleration can be calculated with Miles' equation

$$3\sigma_a = 3\sqrt{\frac{\pi}{2} f_n Q W_{\ddot{U}}(f_n)} = 31.7331 \text{ g}. \tag{11.50}$$

The bending stress σ_b at the root of the canister is

$$\sigma_b = M_b/W_s = 9.81(0.5mL^2 + ML)3\sigma_a/W_s = 32.4029 \text{ MPa}, \tag{11.51}$$

where W_s is the bending resistance $W_s = I/R$.

The shear force D at the root of the canister is given by

$$D = 9.81(mL + M)3\sigma_a \tag{11.52}$$

Subsequently with aid of Jourawski equation, the shear stress at the root of the canister can be calculated

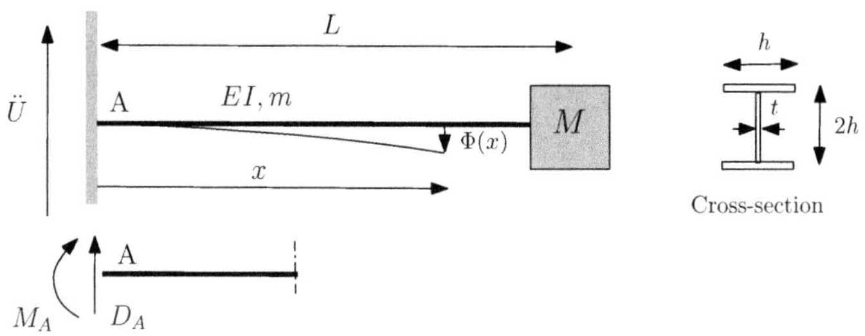

Fig. 11.18 Cantilevered beam with tip mass

$$\tau = \frac{DS}{2It} = 609.4759\,\text{MPa}, \tag{11.53}$$

where $S \approx 2R^2t$ is the first moment of area with respect to the neutral line, t is the wall thickness of the cylinder, and $I \approx \pi R^3 t$ is the second moment of area. The shear stress τ is rather high.

11.7 Because of a presentation of J. Brent Knight of NASA MSFC about conservatism in random vibration design loads at the 'Spacecraft and Launch Vehicle Environments Workshop', June 19–21, 2012, this artificial problem is created and worked out. It turned out that the higher the natural frequencies of structures the lower are the internal loads, e.g., stress, strain. The cantilevered beam with a mass M at the tip is excited by a random enforced acceleration $\ddot{U}$. Further design parameters are shown in Fig. 11.18. The relative deflection of the beam $W(x, t)$ is divided into the time-dependent generalized coordinate $\eta(t)$ and the assumed mode $\Phi(x)$, such that

$$W(x, t) = \eta(t)\Phi(x). \tag{11.54}$$

The cantilevered beam and tip mass are transformed into a SDOF system given by

$$\ddot{\eta}(t) + 2j\zeta\omega_n\dot{\eta}(t) + \omega_n^2\eta(t) = \Gamma\ddot{U}, \tag{11.55}$$

where

- The assumed mode $\Phi(x) = x^2\left(6L^2 - 4xL + x^2\right)$.
- ζ is the damping ratio, $Q = 1/2\zeta$.
- ω_n $(f_n = \omega_n/2\pi)$ the natural frequency, $\omega_n^2 = \frac{1296EI}{405L^3M+104mL^4}$.
- Γ is the modal participation factor, $\Gamma = -\frac{135M+54mL}{405L^4M+104m\cdot L^5}$.

Try to recapitulate the variables ω_n and Γ.

The r.m.s. value of the generalized coordinate η_{rms} can be calculated with

$$\eta_{rms}(f_n) = \frac{\Gamma}{(2\pi f_n)^2}\sqrt{\frac{\pi}{2}f_nQW_{\ddot{U}}(f_n)} = \frac{\Gamma}{(2\pi f_n)^{1.5}}\sqrt{\frac{Q}{4}W_{\ddot{U}}(f_n)}. \tag{11.56}$$

The r.m.s. value of the physical displacement $W(x,t)$ can now obtained by the following expression

$$w(x)_{rms} = \eta_{rms}\Phi(x). \tag{11.57}$$

The r.m.s value of the bending moment at the root of the beam M_A is given by

$$M_{A,rms} = EI\frac{d^2 w_{rms}}{dx^2}\Big|_{x=0} = 12EIL^2\eta_{rms}, \tag{11.58}$$

and the r.m.s value of the shear force at the root of the beam D_A is given by

$$D_{A,rms} = EI\frac{d^3 w_{rms}}{dx^3}\Big|_{x=0} = 24EIL\eta_{rms}. \tag{11.59}$$

From Eq. (11.56), it may be concluded that the r.m.s. values of the bending moment $M_{A,rms}$ and the shear force $D_{A,rms}$ are strongly dependent on the natural frequency f_n and decrease proportional with a factor $\frac{1}{(2\pi f_n)^{1.5}}$ assuming that $\sqrt{\frac{Q}{4}W_{\ddot{U}}(f_n)}$ is more or less constant.

Numerical values are now introduced: $h = 50$ mm, $t = 1$ mm, $L = 0.6$ m, $E = 70$ GPa, $\rho = 2700$ kg/m^3. The tip mass M varies between 0 and 10 kg. The white noise random enforced acceleration PSD is $W_{\ddot{U}} = 0.04$ g^2/Hz and $g = 9.81$ m/s^2. The area of the cross section of the beam is $A = 4ht$ m^2 and the second moment of

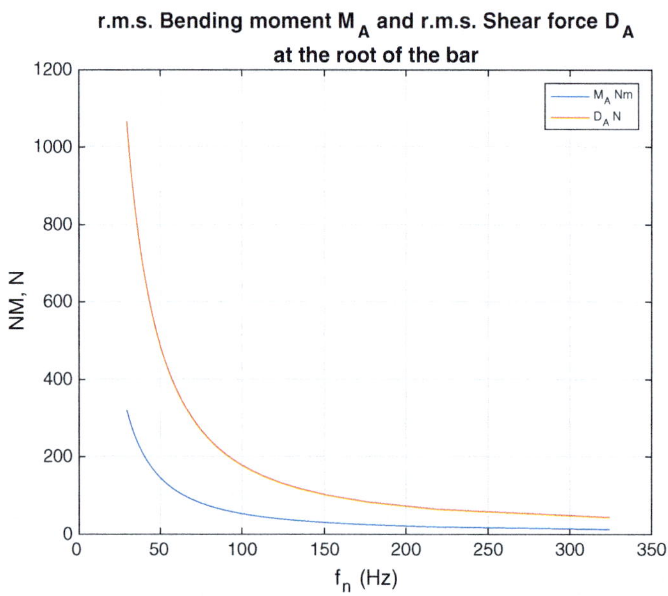

Fig. 11.19 r.m.s. values of $M_A(f_n)$ and $D_A(f_n)$

area is $I = \frac{8}{3}h^3 t$ m^4. The amplification factor is $Q = 10$. The r.m.s. values of the bending moment $M_A(f_n)$ Nm and the shear force $D_A(f_n)$ N are calculated and shown in Fig. 11.19. Beyond $fn > 100$ Hz the r.m.s. value of D_A goes already asymptotic to zero and the r.m.s. value of M_A (strain) goes beyond $f_n > 250-300$ Hz to zero.

References

1. den Hartog JP (1961) Strength of materials. Dover Publications Inc., New York
2. Wilson EL, Penzien J (1972) Evaluation of orthogonal damping matrices. Int J Numer Methods Eng 4(1):5–10
3. D'Souza A, Garg VK (1984) Advanced dynamics, modeling and analysis. Prentice Hall, Englewood Cliffs. ISBN 01-13-011312-3
4. Shi SB, Shao WS, Wei XK, Yang XS, Wang BZ (2016) A new uncondiotional stable fdtd method based on the newman-beta algoritm. IEEE Trans Microw Theory Tech 64(12):4082–4090. https://doi.org/10.11109/TMTT.2016.2608340
5. Wood WL (1990) Practical time-stepping schemes. Clarendon Press, Oxford. ISBN 0-19-859677
6. de la Fuente E (2008) An efficient procedure to obtain exact solutions in random vibration analysis of linear structures. Eng Struct 30:2981–2990
7. Gersch W (1969) Mean-square responses in structural systems. J Acoust Soc Am 48(1, Part 2)
8. Miles RN (1992) Effect of spectral shape on acoustic fatigue life estimation. J Sound Vib 153(2):376–386
9. Meirovitch L (1975) Elements of vibration analysis, International student edn. McGraw-Hill, New York. ISBN 0-07-041340-1
10. Wijker JJ (2009) Random vibrations in spacecraft structures design, theory and applications. Number SMIA 165 in solid mechanics and its applications. Springer, Berlin. ISBN 978-90-481-2727-6
11. Temple G, Bickley WG (1956) Rayleigh's principle and applications to engineering. Dover Publications, New York
12. Himmelblau H, Kern DL, Manning JE, Piersol AG, Rubin S (2001) Dynamic environmental criteria, volume NASA-HDBK-7005, NASA
13. Trubert M (1989) Mass acceleration curve for spacecraft structural design. Report JPL D-5882, NASA, Jet Propulsion Labatory
14. Steinberg DS (2000) Vibration analysis for electronic equipment, 3rd edn. Wiley, New York. ISBN 0-471-37685-X
15. Kwon YW, Bang H (2000) The finite element method, using MATLAB, 2nd edn. CRC Press, Boca Raton. ISBN 0-8493-0096-7
16. Stephen NG (1982) Note on the combined use of dubkerley's and southwell's methods. J Sound Vib 83(4):585–587
17. Wijker JJ (2004) Mechanical vibrations in spacecraft design. Springer, Berlin. ISBN 3-540-40530-5

Appendix A
Random Response SDOF System

In this appendix, the random response of a single degree of freedom (SDOF) system, excited by a random enforced acceleration, is explained. The spectral approach is used. This appendix is added to this book because in a number of chapters this method is applied to compute the power spectral density (PSD) and the root mean square (r.m.s.) value of the response of the mass of the SDOF system. The SDOF system is shown in Fig. A.1.

The equation of motion of the SDOF system is given by

$$m\ddot{X}(t) + c\dot{X}(t) + kX(t) = c\dot{U}(t) + kU(t), \tag{A.1}$$

where all variables are denoted in Fig. A.1. When the variables in the time domain are transformed into the frequency domain, Eq. (A.1) can be written as follows

$$\ddot{X}(\omega) + 2j\zeta\omega_n\dot{X}(\omega) + \omega_n^2X(\omega) = \left[-2j\zeta\frac{\omega_n}{\omega} - \left(\frac{\omega_n}{\omega}\right)^2\right]\ddot{U}(\omega), \tag{A.2}$$

where ζ is the damping ratio, ω_n the natural frequency, and ω the excitation frequency. The acceleration $\ddot{X}(\omega)$ can be expressed in the enforced acceleration $\ddot{U}(\omega)$.

$$\ddot{X}(\omega) = \left[\frac{2j\zeta\left(\frac{\omega_n}{\omega}\right) + 1}{1 - \left(\frac{\omega_n}{\omega}\right)^2 + 2j\zeta\left(\frac{\omega_n}{\omega}\right)}\right]\ddot{U}(\omega) = H(j\omega)\ddot{U}(\omega), \tag{A.3}$$

where $H(j\omega)$ is the frequency transfer function (FRF).

The PSD of the acceleration $W_{\ddot{X}}$ can be expressed in the PSD of the enforced acceleration $W_{\ddot{U}}$

$$W_{\ddot{X}}(f) = |H(f)|^2W_{\ddot{U}}(f), \tag{A.4}$$

J. Wijker, *Miles' Equation in Random Vibrations*, Solid Mechanics and Its Applications 248, https://doi.org/10.1007/978-3-319-73114-8

Fig. A.1 SDOF system, enforced acceleration

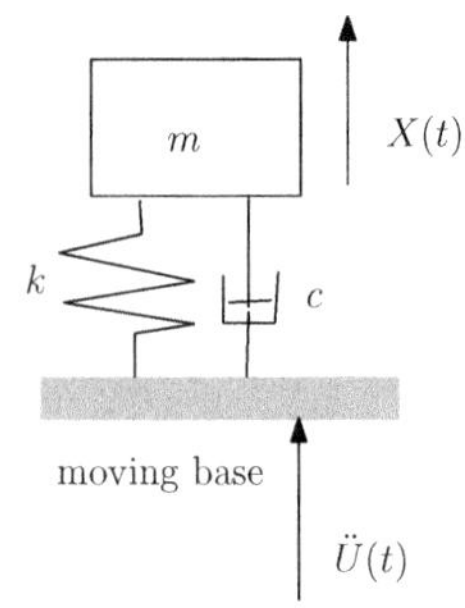

and

$$H(jf) = \left[\frac{2j\zeta\left(\frac{f_n}{f}\right) + 1}{1 - \left(\frac{f_n}{f}\right)^2 + 2j\zeta\left(\frac{f_n}{f}\right)} \right], \tag{A.5}$$

where the rad/s are replaced by cycles/s (Hz).

The r.m.s. value of the acceleration $\ddot{X}$ can be calculated using

$$\ddot{x}_{rms} = \int_0^\infty W_{\ddot{X}}(f)df. \tag{A.6}$$

The integral in Eq. (A.6) can be approximated (numerical quadrature) by the trapezoidal method to calculate $\ddot{x}_{rms}$ numerically [1].

$$\ddot{x}_{rms}(f_n) = \sqrt{\sum_{k=1}^{N} |H(f_n, f_k)|^2 W_{\ddot{X}}(f_k)\Delta f_k}, \tag{A.7}$$

where Δf_k is the step size (increment) between two subsequent frequency steps.

Example

Given the random acceleration specification in Table A.1 [2]. Calculate the random response spectrum (RRS) of that acceleration specification for an instrument with mass $m = 10$ kg and an amplification factor $Q = 10$. The RRS is the r.m.s. response

Table A.1 Specifications for random testing [2]

Frequency (Hz)	PSD (g^2/Hz)
20–100	+3 dB/oct.
100–400	$0.05 \times \frac{m+20}{m+1}$
400–2000	−3 dB/oct.
m (kg)	Mass equipment with unknown location

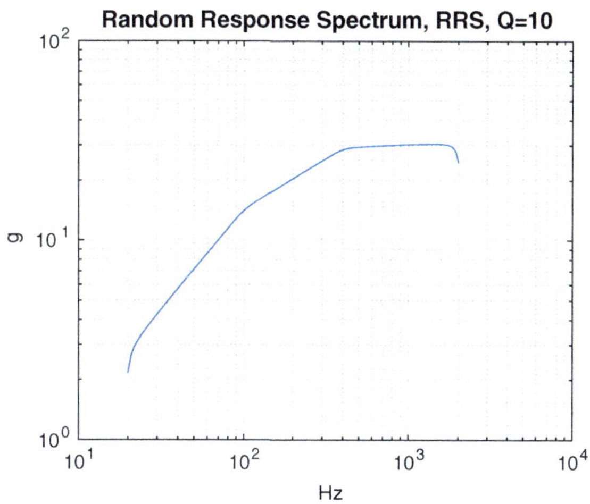

Fig. A.2 RSS of random acceleration specification Table A.1

of a SDOF system to the random excitation as function of the natural frequency and for a given damping ratio ζ. The RRS is presented in Fig. A.2. The RRS may be considered to reflect the damaging effect of the environment.

The PSD of the random acceleration spectrum can be reconstructed from the RRS using Miles' equation

$$W_{\ddot{U}}(f_n) = \frac{RRS^2(f_n)}{\frac{\pi}{2} f_n Q}. \tag{A.8}$$

Problems

A.1 An given RRS has a logarithmic varying acceleration of 10 g at 20 Hz and 300 g at 2000 Hz. The frequency varies linearly. Calculate the corresponding PSD of the acceleration specification with an amplification factor $Q = 10$.

A.2 Calculate the RSS from the PSD of the acceleration spectrum computed in Problem A.1 with a damping ratio $\zeta = 0.05$.

Table A.2 Random acceleration specification $W_{\ddot{U}}$

Spectrum 1		Spectrum 2	
Frequency (Hz)	(g^2/Hz)	Frequency (Hz)	(g^2/Hz)
20–100	3 dB/oct		
100–500	0.04	20–2000	0.01
500–2000	−6 dB/oct		

Table A.3 Random vibration specification $W_{\ddot{U}}$

Frequency (Hz)	PSD random acceleration (g^2/Hz)
20–80	3 dB/oct.
80–500	0.04
500–2000	−6 dB/oct.

A.3 In Table A.2 two spectra of the enforced random acceleration $W_{\ddot{U}}$ are specified. Evaluate the potential damaging characteristics of both spectra using the RSS (Q = 10).

A.4 Calculate the extreme response spectrum $ERS(f)$ of the random enforced acceleration specification given in Table A.3 using Miles' equation. The ERS is an equivalent of the shock response spectrum (SRS).

Calculate G_{rms} and perform the following steps:

- Calculate $ERS(f) = n\sqrt{\frac{\pi}{2} fQW_{\ddot{U}}(f)}$, with f the frequency between 20 and 2000 Hz and $Q = 10$.
- $n = \sqrt{2\ln(fT)}$ is a (peak) factor on the standard deviation to yield maximum response. The test duration is $T = 120$ s.

Answer: $G_{rms} = 6.78$ g

References

1. Schwarz HR (1989) Numerical analysis. Wiley, New York. ISBN 0-471-92065-7
2. Girard A, Imbert JF, Moreau D (1989) Derivation of European satellite equipment test specification from vibro-acoustic test data. Acta Astronaut 10(10):797–803

Appendix B
Quasi-static, Random, and Acoustic Loads

B.1 Introduction

During flight, the spacecraft is subjected to static and dynamic loads. Within the frame of this book about Miles' equation, only the following static and dynamic loads are discussed:

- Quasi-static load.
- Acoustic loads.
- Random vibrations.

The specifications about the quasi-static design loads, the manner how the acoustics loads and the random vibrations are prescribed need more explanation, which will be given in the following sections.

B.2 Quasi-static Load Specifications

Figure B.1 shows a typical longitudinal static acceleration time history for the Soyuz launch vehicle during its ascent flight. The highest longitudinal acceleration occurs just before the first stage cutoff and does not exceed 4.3 g. The highest lateral static acceleration may be up to 0.4 g at maximum dynamic pressure and takes into account the effect of wind and gust encountered in this phase [1]. The very low frequency dynamic loads, which are expected to have no influence on the dynamic behavior of the spacecraft, are added to the static acceleration. The static acceleration increased by the low frequency dynamic loads is called the quasi-static loads (QSL).

The quasi-static design loads are multiplied by the factors of safety to construct design limit loads to design the spacecraft structure and to analyze strength and stiffness characteristics of the spacecraft structure such as: permanent deformation, ultimate strength, buckling.

J. Wijker, *Miles' Equation in Random Vibrations*, Solid Mechanics
and Its Applications 248, https://doi.org/10.1007/978-3-319-73114-8

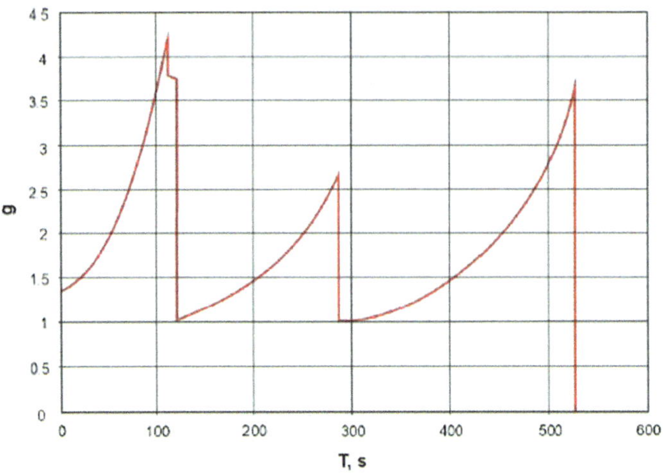

Fig. B.1 Typical longitudinal static acceleration [1]

B.3 Acoustic and Random Load Specification

B.3.1 Introduction

The acoustic noise inside the fairing (shroud) of the launch vehicle is assumed to be reverberant (diffuse); however, this is a mathematical representation of the real sound field. The reverberant sound field representation is well accepted in the spacecraft structures design community. This sound field under the fairing will excite the outer structure of the spacecraft as well as the subsystems mounted to the outside of the spacecraft, such as solar array wing, antennae dishes. The random sound pressures exciting the outer structure of the spacecraft are transferred into random structural vibrations and acoustic pressures of inside cavities. The interior acoustic noise will excite the internal spacecraft structure and is transferred to random structural vibrations too. This process is illustrated in Fig. B.2. The random structural vibrations will excite panel mounted equipment and components at the base. In general, small equipment is not sensitive to acoustic noise.

The typical representation of acoustic load and random acceleration specifications is discussed in subsequent sections [2, 3].

B.3.2 Acoustic Loads

Acoustic loads appear as design specifications for spacecraft and spacecraft attachments such as solar arrays and antennae. Acoustic loads are generated during launch,

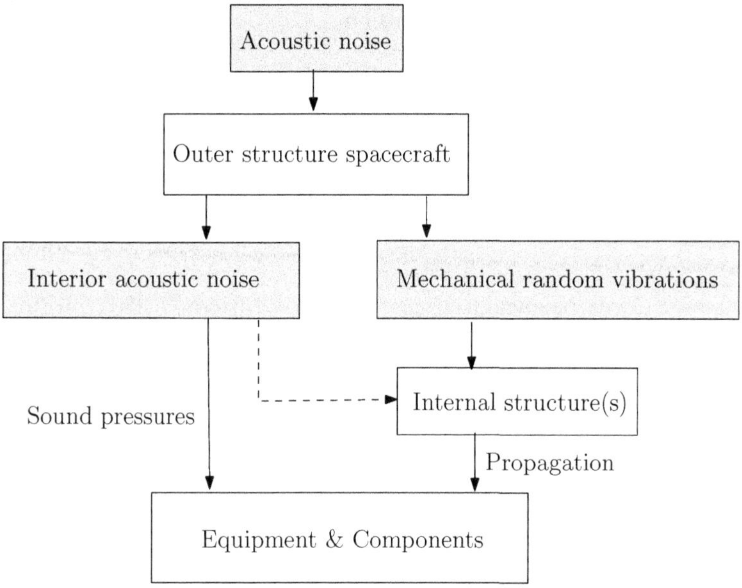

Fig. B.2 Mechanism to produce random vibrations

or in acoustic facilities for test purposes, e.g., reverberation chamber.[1] It is very common to specify a reverberant sound field [1], which means that the intensity of the sound is the same for all directions. More about the fundamentals of acoustics or sound can be read in [4–6].

In general, the acoustic loads are described as sound pressure levels (SPL) and specified in decibels (dB) [1, 7]. The SPL is defined by

$$SPL = 10log\left(\frac{p}{p_{ref}}\right)^2, \tag{B.1}$$

where p is the rms pressure in a certain frequency band with frequency bandwidth Δf, mostly one octave band or one-third octave band, and p_{ref} is the reference pressure 2×10^{-5} Pa.

The xth octave band of two sequential frequencies f_1 and f_2 is given by

$$\frac{f_2}{f_1} = 2^x, \tag{B.2}$$

where $x = 1$ for the octave, and $x = \frac{1}{3}$ when the one-third octave band; then $\frac{f_2}{f_1} = 2^{\frac{1}{3}} = 1.260$.

[1] A reverberation chamber is an enclosure with thick, rigid walls and smooth interior surfaces that strongly reflect sound waves.

The center frequency f_c (Hz) is defined by

$$f_c = \sqrt{f_{min} f_{max}}, \tag{B.3}$$

where f_{min} (f_{max}) is the minimum (maximum) frequency (Hz).

The frequency bandwidth Δf (Hz) is given by

$$\Delta f = f_{max} - f_{min}. \tag{B.4}$$

With $\frac{f_{max}}{f_{min}} = 2^x$, the bandwidth Δf can be expressed in terms of the center frequency f_c as follows

$$\Delta f = \left(2^{\frac{x}{2}} - 2^{\frac{-x}{2}}\right) f_c. \tag{B.5}$$

When

- $x = 1$, the one octave band width is $\Delta f = 0.7071 f_c$.
- $x = \frac{1}{3}$, the one-third octave band width is $\Delta f = 0.2316 f_c$.

The PSD of the pressure field $W_p(f_c)$ (Pa^2/Hz) in the frequency band with center frequency f_c, bandwidth Δf, and r.m.s. pressure $p(f_c)$ is defined as

$$W_p(f_c) = \frac{p^2(f_c)}{\Delta f(f_c)}. \tag{B.6}$$

The mean square value p^2_{rms} of the sound pressure level is given by

$$p^2_{rms} = \int_0^\infty W(f) df = \sum_{i=1}^{k} W_i(f_c) \Delta f = \sum_{i=1}^{k} p_i^2(f_c), \tag{B.7}$$

where k is the number of one octave or one-third octave bands.

The overall sound pressure level (OASPL) in dB is defined as

$$OASPL = 10 log\left(\frac{p^2_{rms}}{p^2_{ref}}\right). \tag{B.8}$$

The following relation determines the conversion of a one-third octave band to a one octave band;

$$SPL_{1-octave} = 10 \log\left[\sum_{k=1}^{3} 10^{\frac{SPL_{\frac{1}{3}-octave}}{10}}\right], \tag{B.9}$$

and the following relation determines the conversion of a one octave band to a one-third octave band

$$SPL_{\frac{1}{3}-octave} = SPL_{1-octave} + 10\log\left[\frac{\Delta f_{\frac{1}{3}-octave}}{\Delta f_{1-octave}}\right]. \tag{B.10}$$

B.3.2.1 Reverberant Sound Field

The cross-power spectral density of the pressure $W_p(f, \mathbf{x})(\mathrm{Pa}^2/\mathrm{Hz})$, in a reverberant (diffuse) sound field (acoustic chamber), at position $\mathbf{x}$ and frequency f (Hz), is given by [8]

$$W_p(f, \mathbf{x}) = W_p(f)\frac{\sin k|\Delta \mathbf{x}|}{k|\Delta \mathbf{x}|}, \tag{B.11}$$

where $W_p(f)(\mathrm{Pa}^2/\mathrm{Hz})$ is the auto-power spectral density of the (reference) pressure p (Pa), $k = 2\pi f/c$ is the wave number, c(m/s) is the speed of sound in air, $|\Delta\mathbf{x}|$(m) the distance between the positions $\mathbf{x}_1$ and $\mathbf{x}_2$. It is stated in [9] that the spectral density between two distinct points $\mathbf{x}_1$ and $\mathbf{x}_2$ in space is:

- dependent on the relative distance $|\Delta\mathbf{x}|$ between two points only, i.e., it is independent of the absolute coordinates of any of the two points. Hence, the sound field is homogenous.
- independent of the direction of the vector $\Delta\mathbf{x}$ between the two points, i.e., it is the same in all directions. Hence, the sound field is isotropic.

Figure B.3 shows a plot of the normalized diffuse (reverberant) field spatial cross-spectral density in air ($c = 340$ m/s) for frequencies $f = 50$, 150, and 500 Hz as a function of the distance Δx. This Fig. B.3 is created using the computer algebra system wxMaxima®, version 11.08.0.

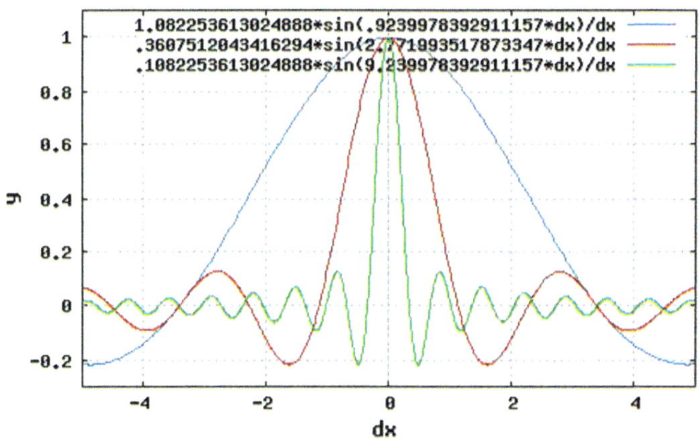

Fig. B.3 Normalized diffuse pressure field cross-spectral density, $y = \sin(k\Delta x)/(k\Delta x)$, $f = 50$ Hz (blue), $f = 150$ Hz (red), $f = 500$ Hz (green), $c = 340$ m/s

A reverberant sound field can be simulated by a combination of (at least 26) plane waves caused by point sources placed equally distributed on a sphere with a large radius [10].

B.3.3 Random Enforced Acceleration

In most cases, the random mechanical loads for spacecraft and subsystems of spacecraft are specified in a very special manner. The power spectral density (PSD) values of the acceleration depend on the frequency (Hz). In general, the frequency range is between 20 and 2000 Hz. The specification must be accompanied by the G_{rms} value of the random acceleration in the frequency range. An example of a typical acceleration specification is given below.

- 20–150 Hz 6 dB/oct
- 150–700 Hz $W_{\ddot{u}} = 0.04$ g^2/Hz
- 700–2000 Hz −3 dB/oct
- $G_{rms} = 7.3$ g.

The graphical representation of the random acceleration specification is shown in Fig. B.4.

The octave band between f_2 and f_1 is defined by

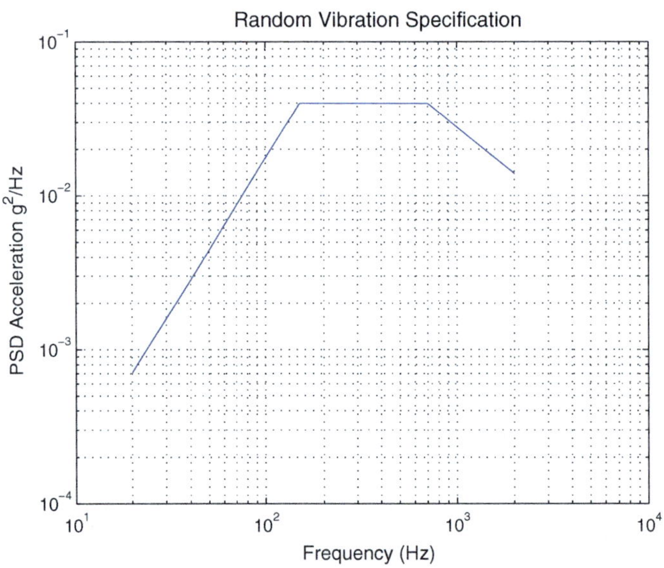

Fig. B.4 Specification PSD acceleration $G_{rms} = 7.3$ g

$$\frac{f_2}{f_1} = 2^1. \tag{B.12}$$

The power of 1 denotes the octave band. The number x is the number of octaves between two frequencies f, and the reference frequency f_{ref} can be obtained using

$$\frac{f}{f_{ref}} = 2^x, \tag{B.13}$$

this yields

$$x = \frac{\ln\left(\frac{f}{f_{ref}}\right)}{\ln 2} = \frac{\log\left(\frac{f}{f_{ref}.}\right)}{\log 2}. \tag{B.14}$$

The relation between the PSD values depends on the number of dBs per octave n (dB/oct) and the number of octaves between two frequencies f and f_{ref}. The relation in dB between $W_{\ddot{u}}(f)$ and $W_{\ddot{u}}(f_{ref})$ is given by

$$10 \log\left\{\frac{W_{\ddot{u}}(f)}{W_{\ddot{u}}(f_{ref})}\right\} = nx = \frac{n \log\left(\frac{f}{f_{ref}}\right)}{\log 2}, \tag{B.15}$$

or

$$\left\{\frac{W_{\ddot{u}}(f)}{W_{\ddot{u}}(f_{ref})}\right\} = \left(\frac{f}{f_{ref}}\right)^{\frac{n}{10 \log 2}} \approx \left(\frac{f}{f_{ref}}\right)^{\frac{n}{3}}. \tag{B.16}$$

If both the frequency f axis and the axis of the PSD function $W(f)$ have a log scale then the angle m (dB/freq) can be obtained by

$$m = \frac{\log W_{\ddot{u}}(f) - \log W_{\ddot{u}}(f_{ref})}{\log f - \log f_{ref}} = \frac{\log\left\{\frac{W_{\ddot{u}}(f)}{W_{\ddot{u}}(f_{ref})}\right\}}{\log\left(\frac{f}{f_{ref}}\right)} = \frac{n}{3}. \tag{B.17}$$

Finally the derivation of the following expression is obtained, a relation between the PSD functions and the frequencies

$$W_{\ddot{u}}(f) = W_{\ddot{u}}(f_{ref})\left(\frac{f}{f_{ref}}\right)^{\frac{n}{3}} = W_{\ddot{u}}(f_{ref})\left(\frac{f}{f_{ref}}\right)^{m}. \tag{B.18}$$

The total root mean square (r.m.s.) value (magnitude) of $\ddot{u}(t)$ is equal to the square root of the area bounded by the PSD function between frequency limits f_1 and f_2. This can be written as

$$\ddot{u}_{rms} = \sqrt{E\left\{\ddot{u}^2(t)\right\}} = \int_{f_1}^{f_2} W_{\ddot{u}}(f) df \tag{B.19}$$

Substituting (B.18) into (B.19) we obtain the following expression for $\ddot{u}_{rms}$

$$\begin{aligned}\ddot{u}_{rms} &= \sqrt{E\left\{\ddot{u}^2(t)\right\}}\\ &= \sqrt{\int_{f_1}^{f_2} W_{\ddot{u}}(f_1)\left(\frac{f}{f_1}\right)^m df} = \sqrt{\frac{W_{\ddot{u}}(f_1)f_1}{m+1}\left[\left(\frac{f_2}{f_1}\right)^{m+1}-1\right]},\quad f_1 < f_2\\ &= \sqrt{-\int_{f_2}^{f_1} W_{\ddot{u}}(f_2)\left(\frac{f}{f_2}\right)^m df} = \sqrt{\frac{W_{\ddot{u}}(f_2)f_2}{m+1}\left[1-\left(\frac{f_1}{f_2}\right)^{m+1}\right]},\quad f_1 < f_2.\end{aligned} \tag{B.20}$$

The parameters needed to calculate the G_{rms} value of the random acceleration spectrum are illustrated in Fig. B.5.

The specification of the PSD (sometimes called acceleration spectral density (AS-D)) of the enforced acceleration or base excitation can be divided into three regions:

- Spectrum with a positive slope n_1 (rising).
- Flat spectrum (slope is zero).
- Spectrum with negative slope n_2 (falling).

B.3.3.1 Spectrum with a Positive Slope

Figure B.5 shows a rising spectrum with a constant slope $n_1 > 0$ between f_1 and f_2. The constant slope is expressed in decibels per octave. The area A_1 can be calculated as follows

$$A_1 = \left\{\frac{W(f_2)f_2}{m_1+1}\left[1-\left(\frac{f_1}{f_2}\right)^{m_1+1}\right]\right\}, \tag{B.21}$$

where $m_1 = n_1/3$, and $n_1 > 0$ is the increase of the PSD value in decibels per octave.

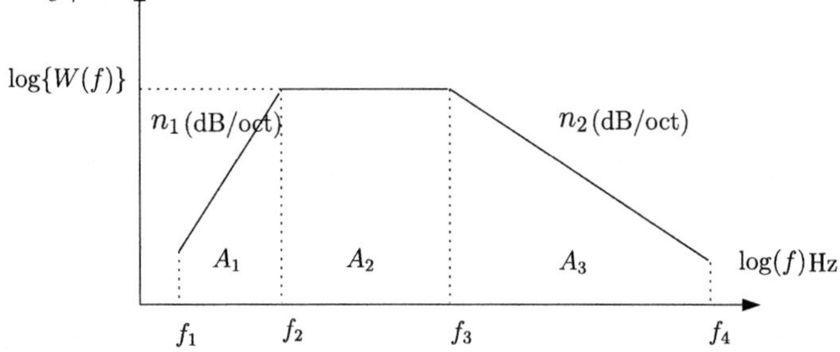

Fig. B.5 Calculation of G_{rms}

B.3.3.2 Flat Spectrum

For a flat spectrum with a zero slope between f_2 and f_3 with $m_1 = 0$ in (B.21), the area A_2 becomes

$$A_2 = W(f_2)\left[f_3 - f_2\right], \tag{B.22}$$

as shown in Fig. B.5.

B.3.3.3 Spectrum with a Negative Slope

For a falling spectrum of a constant slope $n_2 < 0$ between f_3 and f_4, the constant slope is expressed in decibels per octave. The area A_3 can be calculated as follows

$$A_3 = \left\{ \frac{W(f_3)f_3}{m_2 + 1} \left[\left(\frac{f_4}{f_3}\right)^{m_2+1} - 1 \right] \right\}, \quad m_2 \neq -1, \tag{B.23}$$

where $m_2 = n_2/3$, and $n_2(< 0)$ is the decrease of the PSD value in decibels per octave. Equation (B.23) is not applicable if $m_2 = -1$. In that case we have to calculate the value of A_3 when $\lim_{m_2 \to -1}$. This limit can be found using L'Hôpital's Rule[2] [11]. If $u(m_2) = \left(\frac{f_4}{f_3}\right)^{m_2+1} - 1$ and $v(m_2) = m_2 + 1$, then L'Hôpital's rule gives

$$A_3 = W(f_3)f_3 \ln\left(\frac{f_4}{f_3}\right) = 2.30W(f_3)f_3 \log\left(\frac{f_4}{f_3}\right), \quad m_2 = -1. \tag{B.24}$$

The G_{rms} of the enforced random acceleration specification can be obtained, as illustrated in Fig. B.5, by the following expression

$$G_{rms} = \sqrt{A_1 + A_2 + A_3}. \tag{B.25}$$

B.4 Random Vibration Test Tolerances

At the end of this appendix, something is said about tolerance on power spectral densities and r.m.s. values of the random enforced accelerations during the random vibration test (Table B.1).

---

[2]Let $u(a) = v(a) = 0$. If there exists a neighborhood of $x = a$ such that (1) $v(x) \neq 0$, except for $x = a$, and (2) $u'(x)$ and $v'(x)$ exist and do not vanish simultaneously, then

$$\lim_{x \to a} \frac{u(x)}{v(x)} = \lim_{x \to a} \frac{u'(x)}{v'(x)}$$

whenever the limit on the right exists.

In the ECSS, testing standard [12] are the allowable tolerances for several kinds of tests specified.

Problems

B.1 Given the following random acceleration specification in Table B.2. Calculate the slopes and G_{rms}.
Answers: 3 dB/oct., −3 dB/oct, $G_{rms} = 6.06$ g.

B.2 Presented in Table B.3 the 'Vega-Users-Manual-issue-04-April-2014' SPL qualification acoustic vibration levels.

Table B.1 Random vibration test tolerances [12]

Frequency range (Hz)	Amplitude PSD/rms
20–1000	$-1/+3$ dB
1000–2000	± 3 dB
Overall G_{rms}	± 10 % g

Table B.2 Random acceleration spectrum

Frequency (Hz)	ASD g^2/Hz	Slopes
20	0.01005	? dB/oct
80	0.04	
350	0.04	
2000	0.00704	? dB/oct
G_{rms}	?	

Table B.3 Acoustic vibration test levels

Octave center frequency (Hz)	SPL qualification levels (dB) $p_{ref} = 2.0 \times 10^{-5}$ Pa
31.5	123
63	126
125	129
250	135
500	138
1000	130
2000	123
OASPL (20–2828 Hz)	140.8
Test duration	120 s

Perform the following assignments:

- Set up the series of frequencies in the one-third octave band.
- Convert the one octave band SPLs in Table B.3 into the one-third octave band.

References

1. Arianespace (2012) Soyuz user's manual, Issue 2, Rev 0 edition. www.arianespace.com
2. Wijker JJ (2008) Spacecraft structures. Springer, Berlin. ISBN 978-3-540-755524
3. Wijker JJ (2009) Random vibrations in spacecraft structures design, theory and applications. Number SMIA 165 in solid mechanics and its applications. Springer, Berlin. ISBN 978-90-481-2727-6
4. Fahy F (1985) Sound and structural vibration, radiation, transmission and response. Academic Press Limited, London. ISBN 0-12-247671-9
5. Norton M, Karczub D (1990) Fundamentals of noise and vibration analysis for engineers, 2nd edn. Cambridge University Press, Cambridge. ISBN 0-251-49913-5
6. Smith PW Jr, Lyon RH (1965) Sound and structural vibration. Technical report NASA CR-160, NASA
7. Arianespace (2011) ARIANE 5 user's Manual, Issue 5, Revision 1 edition
8. Waterhouse RV, Cook RK, Berendt RD, Edelman S, Thompson MC (1955) Measurement of correlation coefficients in reverberant sound fields. J Acoust Soc Am 27(6):1072–1077
9. Witting M (1986) Modelling of diffuse sound field and dynamic response analysis of lightweight structures. PhD thesis, Technische Universität München, Hert Utz Verlag. ISBN 3-89675-678-8
10. Hamdi MA, Gardner B (2007) Vibro-acoustic analysis of space structures: state of the art and analysis guidelines. ESA Contract 20371/NL/SFe VASI-ESI-TN-010, ESA
11. Korn GA, Korn TM (1961) Mathematical handbook for scientists and engineers. McGraw-Hill Book Company, New York
12. European Cooperation of Space Engineering, Noordwijk, The Netherlands. Space engineering testing, ECSS-E-ST-10-03C, 3rd edn, 1st June 2012

Appendix C
Simulation of the Random Time Series

C.1 Introduction

In this appendix, the transformation of a PSD function in a random time series with the FFT is discussed. The back transformation of random time series into PSD using IFFT is discussed too.

C.2 Random Time Series

The simulation of random time series is discussed in [1–3]. In this appendix, we discuss the efficiency of fast computation of the random time series. The analysis is based on [2]. If $\Phi_x^1(\omega)$ is the one-sided PSD of the desired signal $x(t)$, then $x(t)$ may approximated by

$$x(t) = \sqrt{2}\sum_{n=0}^{N-1}[\Phi_x^1(\omega_n)\Delta\omega]^{\frac{1}{2}}\cos(\omega_n t - \phi_n), \tag{C.1}$$

where ϕ_n are uniformly distributed random numbers on the interval $(0 - 2\pi)$ and $\omega_n = n\Delta\omega$ and $\Delta\omega = \omega_{max}/N$. ω_{max} is the maximum frequency in the power spectrum $\Phi_x^1(\omega)$, and N is the total number of terms in the summation.

A considerable improvement in the computational effort can be obtained by re-casting (C.1) to allow the use of the fast Fourier transform (FFT). To accomplish this, (C.1) may be written as

$$x(t) = \Re\left[\sqrt{2}\sum_{n=0}^{N-1}[\Phi_x^1(\omega_n)\Delta\omega]^{\frac{1}{2}}e^{j(\omega_n t-\phi_n)}\right]. \tag{C.2}$$

J. Wijker, *Miles' Equation in Random Vibrations*, Solid Mechanics and Its Applications 248, https://doi.org/10.1007/978-3-319-73114-8

If the simulated time series $x(t)$ is needed only at discrete values of time t, let $x_n = x(t_k) = x(k\Delta t)$, where the time duration between the equally spaced samples is Δt. Evaluation of (C.2) at time $t = t_k$ gives

$$x(t_k) = x(k\Delta t) = \Re\left[\sqrt{2}\sum_{n=0}^{N-1}[\Phi_x^1(\omega_n)\Delta\omega]^{\frac{1}{2}} e^{j(\omega_n k\Delta t-\phi_n)}\right]. \qquad (C.3)$$

To satisfy the Nyquist sampling criterion, the time series, $x(t)$, must be sampled at a high enough rate to obtain two samples during one period of the highest frequency component ω_{max} of interest in the original PSD $\Phi_1(\omega)$. Hence, Δt is chosen to be

$$\Delta t \leq \frac{\pi}{\omega_{max}}. \qquad (C.4)$$

The term $\omega_n k\Delta t$ in (C.3) can be rewritten as follows

$$\omega_n k\Delta t = n\Delta\omega k\Delta t = n\frac{\omega_{max}}{N}k\frac{\pi}{\omega_{max}} = \frac{nk2\pi}{2N}. \qquad (C.5)$$

Thus, (C.3) can expressed in the following form

$$x(k\Delta t) = \Re\left[\sqrt{2}\sum_{n=0}^{N-1}[\Phi_x^1(\omega_n)\Delta\omega]^{\frac{1}{2}} e^{-j\phi_n} e^{\frac{jnk2\pi}{2N}}\right]. \qquad (C.6)$$

Equation (C.6) may be evaluated using the FFT algorithm. Given a discrete sequence a_n, the FFT provides an efficient means of computing A_k, where

$$A_k = \sum_{n=0}^{N-1} a_n e^{\frac{jnk2\pi}{2N}}, \quad k = 0, 1, 2, \ldots, N-1. \qquad (C.7)$$

Equation (C.6) may be evaluated using the FFT by defining a sequence

$$\begin{aligned} a_n &= [\Phi_x^1(\omega)\Delta\omega]^{\frac{1}{2}} e^{-j\phi_n} \quad n \leq N-1, \\ &= 0, \quad n \geq N. \end{aligned} \qquad (C.8)$$

Equation (C.6) may then be written as

$$x_k = x(k\Delta t) = \Re\left[\sqrt{2}\sum_{n=0}^{2N-1} a_n e^{\frac{jnk2\pi}{2N}}\right] \quad k = 0, 1, 2, \ldots, 2N-1. \qquad (C.9)$$

Table C.1 Random vibration specification [4]

Frequency (Hz)	W(f) g^2/Hz
20–100	3 dB/oct
100	0.2
600	0.2
600–2000	−3 dB/oct
Overall	15.94 Grms

In finding the real part of (C.9), we may use the complex conjugate[3] of the right side of (C.9) to give

$$x_k = x(k\Delta t) = \Re\left[\sqrt{2}\sum_{n=0}^{2N-1} a_n e^{\frac{-jnk2\pi}{2N}}\right] \quad k = 0, 1, 2, \ldots, 2N-1. \tag{C.10}$$

This is equivalent to

$$x_k = \sqrt{2}\Re[FFT(a_n)]. \tag{C.11}$$

Note that the length of the sequence a_n is $2N$.

The PSD function $\Phi_x^1(\omega)$ can be reconstructed using the following expression

$$\Phi_{x,n}^1 = 2\frac{|IFFT(x_k)|^2}{\Delta\omega} \quad k = 0, 1, 2, \ldots, 2N-1, \quad n = 0, 1, 2, \ldots, N-1, \tag{C.12}$$

where $IFFT$ is the inverse fast Fourier transform.

If the product $\Phi_x^1(\omega)\Delta\omega$ is replaced by $W_x^1(f)\Delta f$, (C.13) can be written as follows

$$W_{x,n}^1 = 2\frac{|IFFT(x_k)|^2}{\Delta f} \quad k = 0, 1, 2, \ldots, 2N-1, \quad n = 0, 1, 2, \ldots, N-1, \tag{C.13}$$

where $W(f)$ is the PSD function in the cyclic frequency domain and f is the cyclic frequency (Hz).

Example

The enforced random acceleration spectrum is specified in Table C.1. This random acceleration spectrum will be simulated by random time series. The number of terms in time series is $N = 500$, and the maximum frequency of interest is $f_{max} = 2000$ Hz. The simulated random time series is shown in Fig. C.1. The reconstructed PSD function $W(f)$ from the random time series in Fig. C.1 is shown in Fig. C.2. The artificial value of the PSD below 20 Hz is $W = 1.0 \times 10^{-6}$ g^2/Hz.

[3] $\cos u = \Re[e^{ju}] = \Re[e^{-ju}]$.

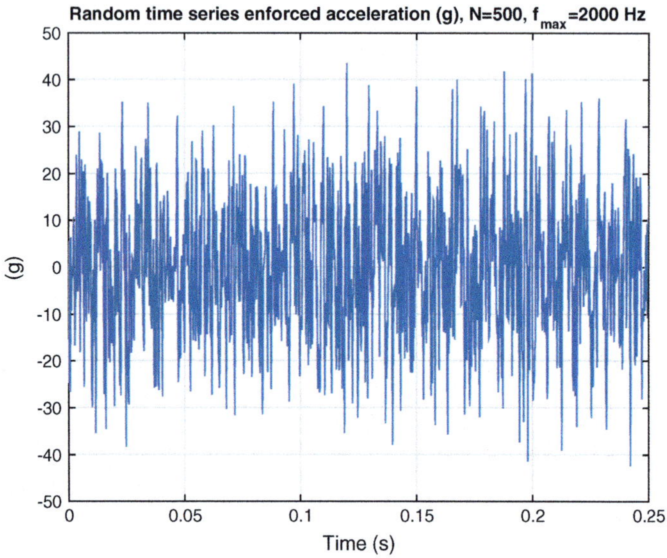

Fig. C.1 Simulated random time series from Table C.1, $1\sigma = 15.95\,g$, $\ddot{U}_{max} = 43.66\,g$

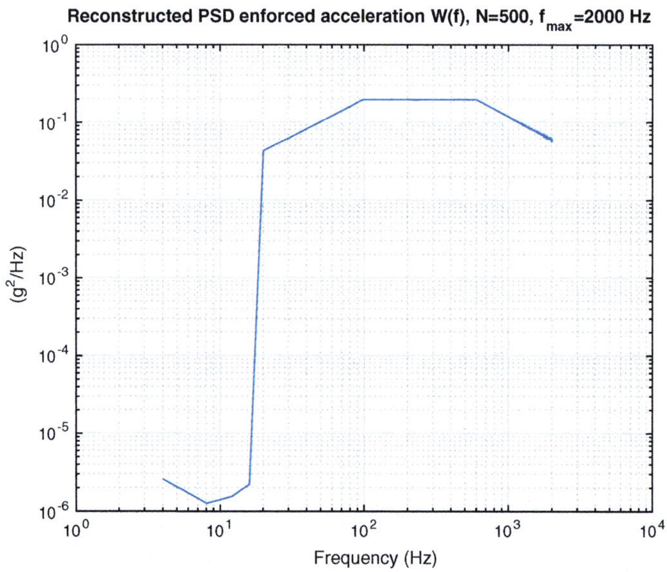

Fig. C.2 Reconstructed PSD $W(f)$ from Fig. C.1

Problems

C.1 This problem is taken from [3]. The PSD function is given by

$$W(f) = W_0 \frac{\alpha}{\alpha^2 + f^2}, \quad 0 \leq f \leq f_u,$$

where

$$W_0 = \left[\arctan \frac{f_u}{\alpha} \right]^{-1},$$

where $f_u = 40\,\text{Hz}$ and $\alpha = 4\,\text{Hz}$.

Perform the following assignments:

- Simulate the random time series.
- Reconstruct the original PSD function from the random time series.

References

1. Cacko J, Bily M, Burkoveecky J (1988) Random processes: measurement, analysis and simulation, vol 8. Fundamental studies in engineering. Elsevier, Amsterdam. ISBN 0-444-98942
2. Miles RN (1992) Effect of spectral shape on acoustic fatigue life estimation. J Sound Vib 153(2):376–386
3. Shinozuka M, Jan CM (1972) Digital simulation of random processes and its applications. J Sound Vib 25(1):111–128
4. Chung YT, Krebs DJ, Peebles JH (2001) Estimation of payload random vibration loads for proper structure design. In: AIAA-2001-1667, Seattle, 42nd AIAA/ASME/ASCE/AHS/ASC structures, structural dynamics, and materials conference 16–19 April 2001

Appendix D
Computation of SRS

D.1 Introduction

The principle of the calculation of the shock response spectrum (SRS) is illustrated in Fig. D.1. For quite a number of SDOF elements, the transient responses are computed when they are excited at the base by a transient acceleration. The maximum values of the transient response $G_k, k = 1, 2, \ldots, n$ of the SDOF element k with natural frequency f_k and amplification factor Q are plotted, successively. The horizontal axis (ordinate) represents the frequency (Hz) and the vertical axis (abscise) the maximum accelerations (g). The figure represents the SRS. The SRS was already mentioned by M.A. Biot in 1933 [1] and later in 1941 [2]. The theoretical description of the SRS was done within the frame of earthquake engineering.

D.2 Single Degree of Freedom System (SDOF)

The equation of motion of the SDOF element (see Fig. D.2) excited at the base by an enforced acceleration $\ddot{u}(t)$ (m/s^2)

$$m\ddot{x} = -c(\dot{x} - \dot{u}) - k(x - u), \tag{D.1}$$

where $\ddot{x}$ (m/s^2) is the absolute acceleration, k(N/m) the stiffness of the spring, and c (Ns/m) the damping of the damper. The transfer function $H(s)$ of this SDOF system in the complex Laplace domain assuming zero initial conditions of the velocity $\dot{x}$ and displacement x is given by

$$H(s) = \frac{\ddot{X}(s)}{\ddot{U}(s)} = \frac{2\zeta\omega_n\omega_n s + \omega_n^2}{s^2 + 2\zeta\omega_n s + \omega_n^2}, \tag{D.2}$$

where the damping ratio is $\zeta = c/2\sqrt{km}$ and the natural frequency is $\omega_n = \sqrt{k/m}$.

J. Wijker, *Miles' Equation in Random Vibrations*, Solid Mechanics and Its Applications 248, https://doi.org/10.1007/978-3-319-73114-8

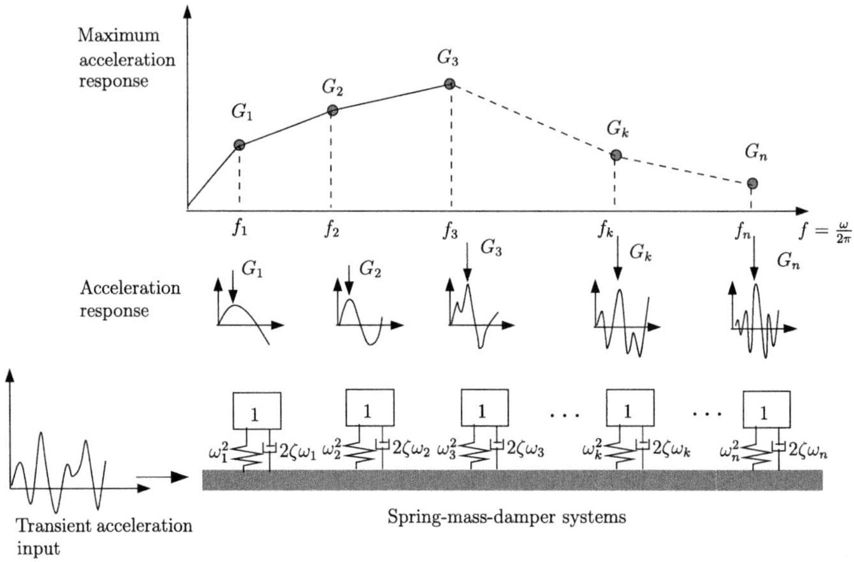

Fig. D.1 Development of a shock response spectrum

Fig. D.2 Base-acceleration-excited SDOF element

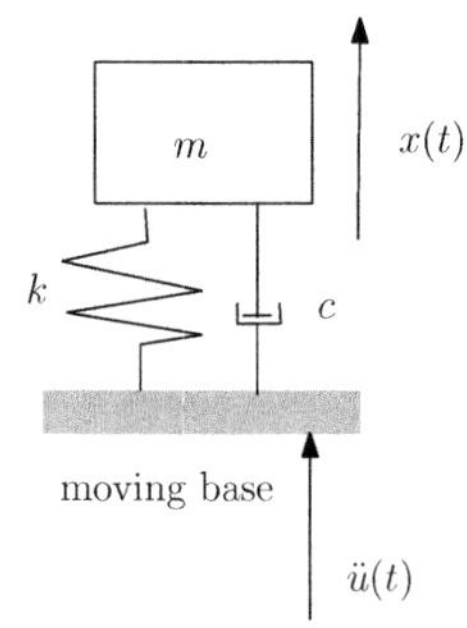

For a given amplification factor $Q = 1/2\zeta$, natural frequency ω_n, and base excitation $\ddot{u}(t)$, it is theoretically possible to calculate the acceleration response $\ddot{x}(t)$ in the form of the time function as well and consequently to determine the peak value of this time function. The input acceleration signal is in the form of a sampled signal with a sample interval $T_s = 1/f_s$, not in the form of a continuous signal. The transfer function Eq. (D.2) has to be approximated by the Z-transform, i.e., to be transformed into a discrete system (digital filter).

D.3 Discrete Approximation of Continuous Transfer Function

The complex function in the s-plane is approximated by a discrete function in the z-plane (z-Transform, [3]), i.e., the approximation of the differential equation of motion by a difference equation, which is a problem of numerical integration [4]. The continuous function between two adjacent samples may be approximated by a constant value or by a ramp or by any suited function. A mapping of the s-plane on the z-plane has to be find saving the important properties of continuous system after transformation into a discrete system of the same order. The transfer function $\tilde{H}(z)$ can be approximated by the First-order hold theorem [3]) using the following expression

$$\tilde{H}(z) = \frac{(z-1)^2}{Tz} Z\left\{L^{-1}\left[\frac{H(s)}{s^2}\right]\right\} \tag{D.3}$$

The digital filter corresponding to the SDOF system responses is the second-order filter infinite impulse response (IIR), with the general z-transform expression [4]:

$$\tilde{H}(z) = \frac{\ddot{X}(z)}{\ddot{U}(z)} = \frac{\beta_0 + \beta_1 z^{-1} + \beta_2 z^{-2}}{1 + \alpha_1 z^{-1} + \alpha_2 z^{-1}}, \tag{D.4}$$

where $\beta_0, \beta_1, \beta_2, \alpha_1$, and α_2 are the filter parameters.

The transfer function $\tilde{H}(z)$ corresponds to the difference equation, which is enabling to calculate the response function in the time domain.

$$\ddot{x}(k) = \beta_0 \ddot{u}(k) + \beta_1 \ddot{u}(k-1) + \beta_2 \ddot{u}(k-2) - \alpha_1 \ddot{x}(k-1) - \alpha_2 \ddot{x}(k-2). \tag{D.5}$$

Equation (D.3) is completely worked out by Irvine in [5]. The digital filter parameters are calculated by equations given below:

$$\begin{aligned}
\beta_0 &= 1 - \exp(-A)\sin(B)/B, \\
\beta_1 &= 2\exp(-A)[\sin(B)/B - \cos(B)], \\
\beta_2 &= \exp(-2A) - \exp(-A)\sin(B)/B, \\
\alpha_1 &= -2\exp(-A)\cos(B), \\
\alpha_2 &= \exp(-2A), \\
A &= \zeta\omega_n T_s = \frac{\omega_n T_s}{2Q}, \\
B &= \omega_n T_s\sqrt{1-\zeta^2} = \omega_n T_s\sqrt{1-\frac{1}{4Q^2}}.
\end{aligned} \tag{D.6}$$

To compute the SRS the following MATLAB™ function can be applied,

```
function [y,f]=srs(x,fs,fmin,fmax,fno,Q)

% SRS calculation using Smallwood improved method
%
%    [y,f]=srs(x,fs,fmin,fmax,fno,Q)
%    y maximum accelerations calculated
%    f frequency vector (Hz)
%    x data vector (enforced acceleration)
%    fs samping frequency (Hz)
%    fmin lowest frequency of interest (Hz)
%    fmax highest frequency of interest (Hz)
%    fno number of frequencies
%    Q Amplification factor
%
y=zeros(fno,1);
f=zeros(fno,1);
%
%
k1=log(fmax/fmin)/(fno-1);
k2=pi/Q/fs;
k3=2*pi/fs*sqrt(1-1/(4*Q^2));
%
%
for n=1:fno;
f0=fmin*exp(k1*(n-1));
A=k2*f0;
B=k3*f0;
a=[1,-2*cos(B)*exp(-A),exp(-2*A)];
b=[1-exp(-A)*sin(B)/B,2*exp(-A)*(sin(B)/B-cos(B)),
exp(-2*A)-exp(-A)*sin(B)/B];
z=filter(b,a,x);
y(n,1)=max(z);
f(n,1)=f0;
end
%endfunction
```

Example

The random response spectrum shown in Fig. D.3 is measured on a solar array wing placed into a reverent acoustic chamber. It is asked to compute a SRS of this given random acceleration spectrum. Before the SRS can be generated, the random acceleration spectrum is transformed into the time domain applying the transformation method discussed in Appendix C. The number of frequency steps is N = 500. The SRS associated with the accelerations in the time domain is shown in Fig. D.5.

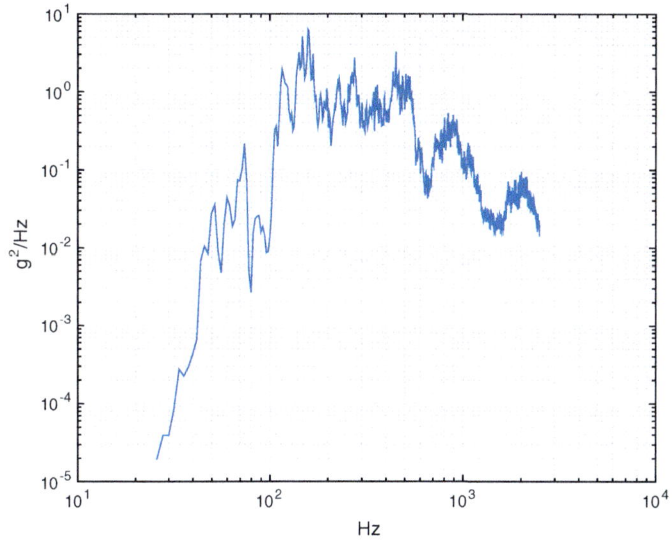

Fig. D.3 Measured random acceleration spectrum, $G_{rms} = 24.98\,\mathrm{g}$

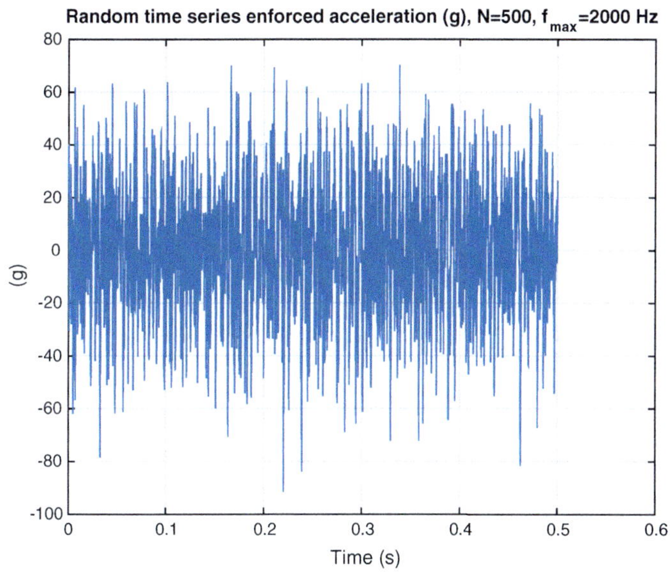

Fig. D.4 Measured random acceleration in time domain, $G_{rms} = 24.98\,\mathrm{g}$

Because the SRS is also based on peak responses, the equivalent random acceleration vibration specification can be obtained using Eq. (10.11) (Fig. D.5).

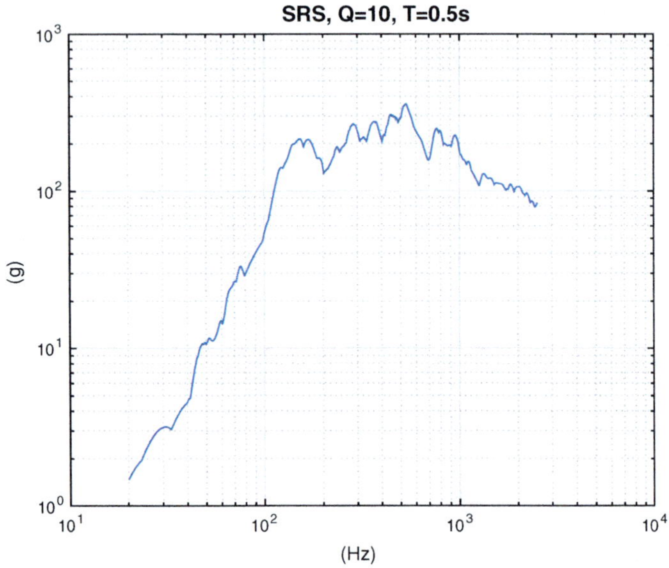

Fig. D.5 Shock response spectrum

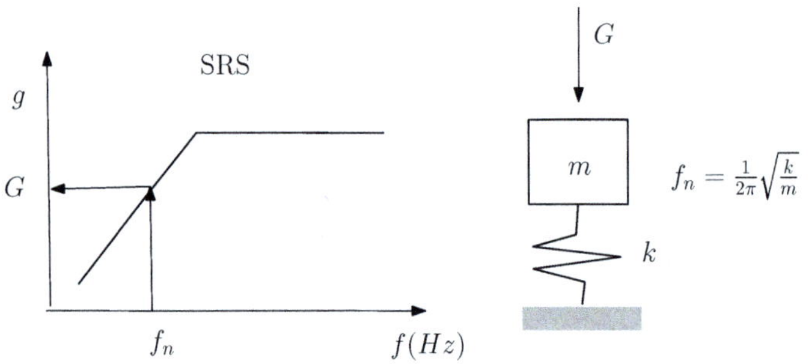

Fig. D.6 Application SRS

Example

In this example, an application of the SRS is presented. In Fig. D.6, the dynamic system (SDOF element) has to be designed to withstand the acceleration G at the natural frequency f_n. The inertia load is mG, and the stress distribution inside the structure can be calculated.

Problems

D.1 This problem is taken from [6]. Find the ramp invariant simulation of

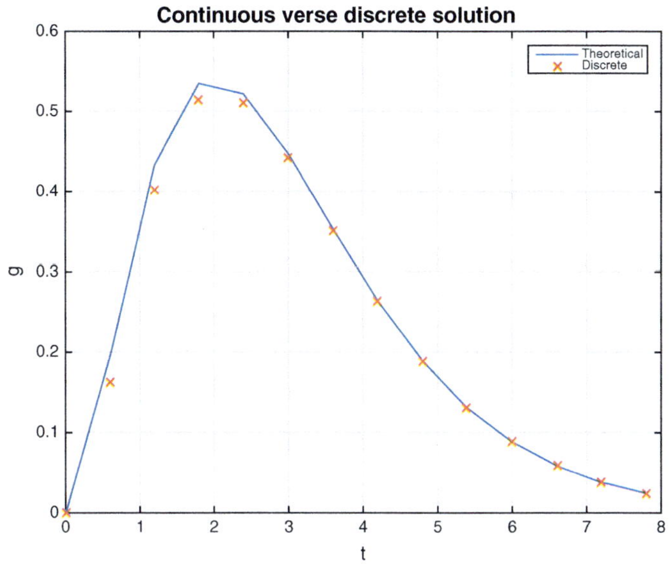

Fig. D.7 Continuous verse discrete solution

$$H(s) = \frac{1}{1+s},$$

and illustrate its response $g(t)$ to $f(t) = 2t\exp(-t)$, $0 \leq t \leq 8$s. $\tilde{H}(z)$ is found as follows:

$$\tilde{H}(z) = \frac{(z-1)^2}{Tz} Z\left\{L^{-1}\left[\frac{H(s)}{s^2}\right]\right\}$$

Define the response function in the time domain. Perform the response analysis in the time domain with $T = 0.6$ s, and compare the numerical results with analytical solution of $g(t) = t^2\exp(-t)$.
Answers:

$$\tilde{H}(z) = \frac{(T-1+\exp(-T))z - T\exp(-T) - 1 - \exp(-T)}{T(z-\exp(-T))},$$

$$g(m) = \frac{1}{T}\left[(T-1+\exp(-T))f(m) - (T\exp(-T) - 1 - \exp(-T)f(m-1)\right] + \exp(_T)g(m-1).$$

Figure D.7.

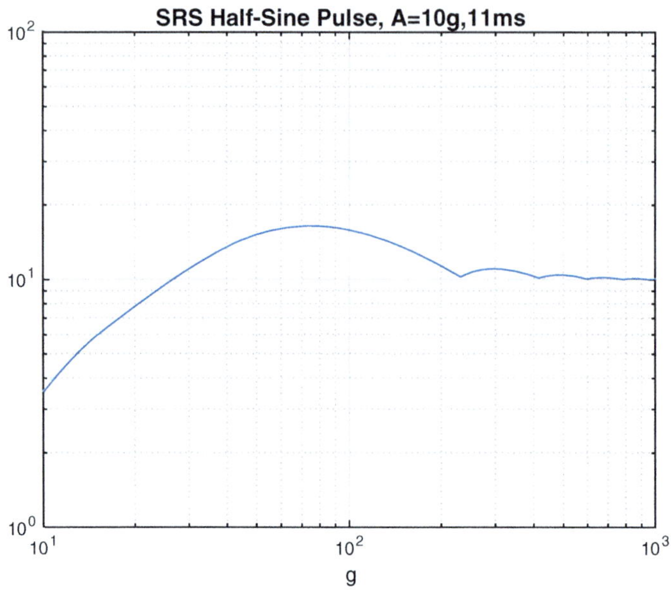

Fig. D.8 SRS half-sine pulse, $A = 10\,\text{g}$, duration 11 ms

D.2 The input acceleration time history is a half-sine pulse of 11 ms is sampled at 2000 Hz [4]

$$\ddot{x} = A\sin(\omega t), 0 \leq t \leq 0.011,$$

where $A = 10$ g.

- Calculate ω.
- Compute the corresponding SRS of $\ddot{x}$ with $Q = 10$.

Answer: $\omega = 1000\pi/11$ rad/s, SRS see Fig. D.8.

D.3 The input acceleration is time history sampled at 2000 Hz which is given by [7]

$$\ddot{x} = u(t)e^{-\eta\omega t}\sin(\omega t) + u(t+\tau)Ae^{-\nu t}\sin(\nu(t+\tau)),\ \ 0 \leq t \leq 0.25,$$

where $u(t)$ is the Heaviside step function, $A = -0.1995$, $\eta = 0.05$, $\omega = 2\pi(100)$, $\nu = 2\pi(10)$, and $\tau = -0.015757$.

- Compute the corresponding SRS of $\ddot{x}$ with $Q = 10$.

References

1. Biot MA (1933) Theory of elastic systems vibrating under transient impulse with an application to earthquake-proof buildings. Proc Natl Acad Sci 19(2):261–268

2. Biot MA (1941) A mechanical analyzer for the prediction of earthquake stresses. Bull Seismol Soc Am 31(2):150–171
3. Stearns SD (2003) Digital signal processing, with examples in matlab. The electrical engineering and applied signal processing series. CRC Press, Boca Raton. ISBN 0-8493-1091-1
4. Tùma J, Koči P (2009) Calculation of shock response spectrum. Colloqium, 2–4 Feb 2009
5. Irvine T (2013) Derivation of the filter coefficients for the ramp invariant method as applied to the base excitation of a single-degree-of-freedom system. www.vibrationdata.com, revision b edition, 3 April 2013
6. Stearns SD (1975) Digital signal analysis. Hayden Book Company, Inc. ISBN 0-8104-5828-4
7. Smallwood DO (1981) An improved recursive formula for calculating shock response spectra. Shock Vib Bull 51:4–10

Appendix E
Application Rayleigh's Quotient

E.1 Introduction

In this appendix, the application Rayleigh's quotient to estimate quickly the natural frequency associated with a particular vibration mode of a dynamic system is discussed.

E.2 Rayleigh's Quotient

The theory behind Rayleigh's quotient or Rayleigh's method can be found in may textbooks, but will be briefly recapitulated in this Appendix. Rayleigh's quotient is defined by the following equation

$$R(u) = \frac{\frac{1}{2}(u)^T[K](u))}{\frac{1}{2}(u)^T[M](u))} = \frac{U}{T^*}, \tag{E.1}$$

where U is the strain energy, T^* the 'kinetic energy', $[K]$ the stiffness matrix, $[M]$ the mass matrix, and (u) the assumed mode. For continuous dynamic systems, the strain energy and 'kinetic energy' can be written as

$$U = \frac{1}{2}\int_V \sigma\varepsilon dV, \tag{E.2}$$

$$T^* = \frac{1}{2}\int_V \rho u dV, \tag{E.3}$$

with σ the stress and ε the strain distribution associated with the assumed displacement u and ρ the density. V is the volume of the continuous system.

Rayleigh's quotient is very useful:

J. Wijker, *Miles' Equation in Random Vibrations*, Solid Mechanics and Its Applications 248, https://doi.org/10.1007/978-3-319-73114-8

- It is insensitive to errors in the choice of the assumed mode.
- It is often possible to guess at the shape of an assumed mode.

Rayleigh's quotient has a stationary value when the assumed mode (u) is in the vicinity of any vibration mode (ϕ_i), with the following well-known orthogonality properties

$$\begin{aligned}(\phi_i)^T[M](\phi_j) &= \delta_{ij}\\(\phi_i)^T[K](\phi_j) &= \omega_i^2\delta_{ij},\end{aligned} \tag{E.4}$$

where δ_{ij} is Kronecker's delta, $\delta_{ii}=1$, $\delta_{ij}=0$, $i\neq j$.

The assumed mode (u) is depicted on an independent set of eigenvectors or vibration modes, such that

$$(u)=\sum_{k=1}^{n}c_k(\phi_k), \tag{E.5}$$

where $c_k,\ k=1,2,\ldots,n$ are the weighting factors. Equation (E.5) is substituted in Eq. (E.1), then

$$(u)^T[K](u))=\sum_{k=1}^{n}c_k^2(\phi_k)^T[K](\phi_k), \tag{E.6}$$

$$(u)^T[M](u))=\sum_{k=1}^{n}c_k^2(\phi_k)^T[M](\phi_k), \tag{E.7}$$

due to the orthogonality relations Eq. (E.4). When $(\phi_k)^T[K](\phi_k)=\omega_k^2(\phi_k)^T[M](\phi_k)$, Rayleigh's quotient can be written

$$R(u)=\omega^2=\frac{\sum_{k=1}^{n}c_k^2\omega_k^2}{\sum_{k=1}^{n}c_k^2}. \tag{E.8}$$

If the assumed mode $(u)\approx(\phi_r)$, then $c_r\gg c_k,\ (r\neq k)$. The Rayleigh quotient can be further evaluated

$$R(u)=\frac{c_r^2\omega_r^2+c_r^2\sum\limits_{\substack{k=1,2,\ldots,\\k\neq r}}\left(\frac{c_k^2}{c_r^2}\right)^2\omega_k^2}{c_r^2+c_r^2\sum\limits_{\substack{k=1,2,\ldots,\\k\neq r}}\left(\frac{c_k^2}{c_r^2}\right)^2}=\frac{\omega_r^2+\sum\limits_{\substack{k=1,2,\ldots,\\k\neq r}}\left(\frac{c_k^2}{c_r^2}\right)^2\omega_k^2}{1+1\sum\limits_{\substack{k=1,2,\ldots,\\k\neq r}}\left(\frac{c_k^2}{c_r^2}\right)^2}. \tag{E.9}$$

The fraction $\left|\frac{c_k^2}{c_r^2}\right|=\varepsilon_k\ll 1$. In case $r=1$

$$R(u) = \frac{\omega_1^2 + \sum\limits_{k=2,3,\dots} \left(\frac{c_k^2}{c_1^2}\right)^2 \omega_k^2}{1 + 1 \sum\limits_{k=2,3,\dots} \left(\frac{c_k^2}{c_1^2}\right)^2}$$

$$\approx \omega_1^2 + \sum_{k=2,3,\dots}^{n} \varepsilon_k^2 \omega_k^2 - \omega_1^2 \sum_{k=2,3,\dots}^{n} \varepsilon_k^2, \tag{E.10}$$

$$= \omega_1^2 + \sum_{k=2,3,\dots}^{n} (\omega_k^2 - \omega_1^2)\varepsilon_k^2.$$

The stationary value of $R(u)$ is

$$\frac{dR}{d\varepsilon} = 0, \text{ if } \varepsilon_k = 0, \ k = 2, 3, \dots \tag{E.11}$$

General conclusions about Eq. (E.10) can now be drawn:

- Rayleigh's quotient is never lower than the first eigenvalue, $R(u) \geq \omega_1^2$, and
- similarly it can be shown that $R(u) \leq \omega_n^2$. Rayleigh's quotient is never higher than the highest eigenvalue.
- In general, Rayleigh's quotient $\omega_1^2 \leq R(u) \leq \omega_n^2$. For continuous dynamic systems, $\omega_n^2 \to \infty$.

Example

A tapered bar is shown in Fig. E.1, with length L, and a varying cross section $A(x) = A_o(1-x/2L)^2, A(0) = A_o$, and $A(L) = A_o/4$. Young's modulus of the bar is E, and the density is ρ. Calculate Rayleigh's quotient when the assumed mode or displacement function is $u(x) = (x/L)^\alpha$.

The strain energy U and kinetic energy T^* of the tapered bar are

$$\begin{aligned} U &= \frac{1}{2}\int_V \sigma\varepsilon dV = \frac{1}{2}\int_0^L EA(x)\left(\frac{du}{dx}\right)^2 dx, \\ T^* &= \frac{1}{2}\int_V \rho u(x)^2 dV = \frac{1}{2}\int_0^L \rho A(x) u(x)^2 dx. \end{aligned} \tag{E.12}$$

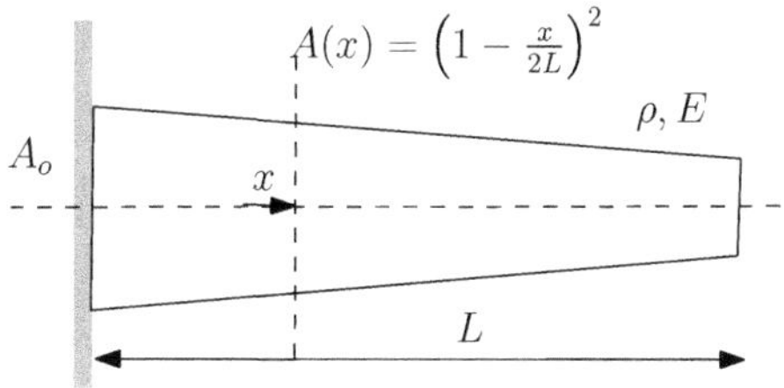

Fig. E.1 A tapered beam

Rayleigh's quotient can now be obtained

$$R(u) = \frac{U}{T*} = \frac{35E}{8\rho L^2}, \tag{E.13}$$

and the lowest approximation of the natural frequency ω is

$$\omega_n = \frac{2.0917\sqrt{\frac{E}{\rho}}}{L} > \omega_{theor} = \frac{2.029\sqrt{\frac{E}{\rho}}}{L}. \tag{E.14}$$

Problems

E.1 A dynamic system consists of three coupled SDOF elements as shown in Fig. E.2. The mass $m = 100$ kg and the spring stiffness $k = 1.0 \times 10^9$ N/m.

Perform the following assignments:

1. Calculate the static displacement vector $(u) = (x_1, x_2, x_3)^T$ under an 1 g gravitational field ($1\text{g} = 9.81\text{m/s}^2$).
2. Calculate Rayleigh's quotient $R(u)$ (Rad/s)^2, and associated natural frequency $f_n(u)$ (Hz).
3. Solve undamped eigenvalue problem $[K] - \omega_i^2[M](\phi_i) = 0$, and subsequently the natural frequencies $f_{n,i} = \omega_i/2\pi,\ i = 1, 2, 3$.

Answers: $(u) = 1. \times 10^{-4}(0.0589, 0.1079, 0.1373)^T$, $R(u) = 8.0831 \times 10^5$(Rad/s)^2, $f_n(u) = 143.0903$ Hz, $(f_n) = (142.7539, 469.1769, 777.0698)^T$ (Hz).

E.2 A cantilevered beam is also supported at the tip by a rotational and a translational spring, k_ϕ and k_w respectively [1]. The system is illustrated in Fig. E.3. Calculate the

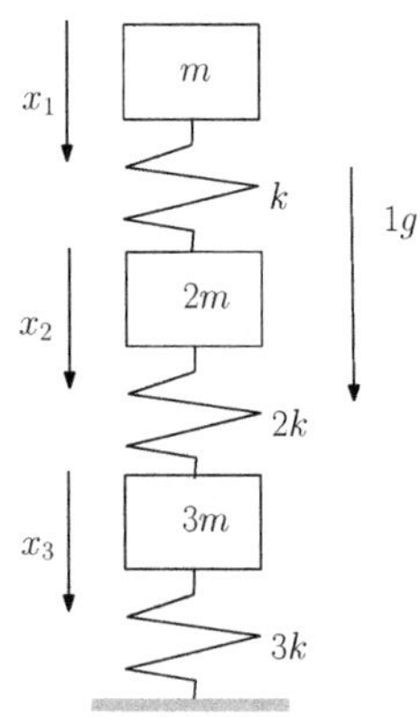

Fig. E.2 Three SDOF dynamic system

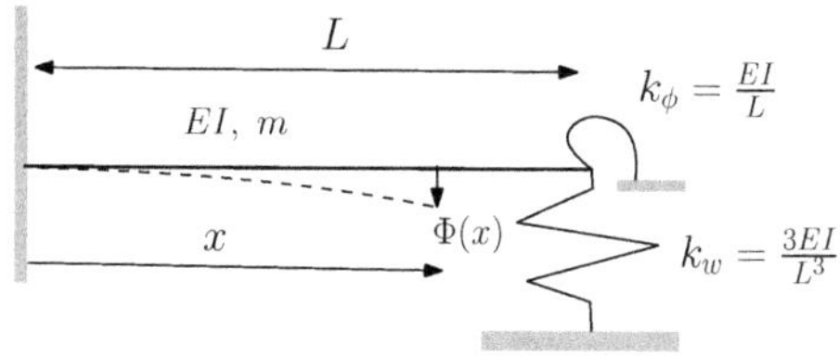

Fig. E.3 Cantilevered beam

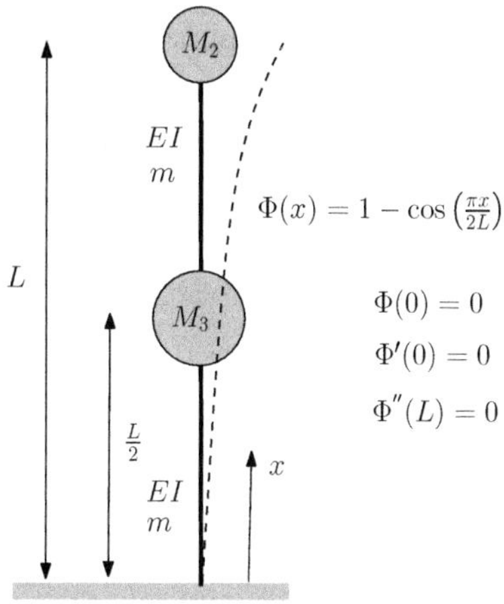

Fig. E.4 S/C modeled as a cantilevered beam

eigenvalue ω_n^2 with the aid of Rayleigh's quotient with the following assumed mode [2, 3]:

$$\Phi(x) = c_1 x^2(6L^2 - 4Lx + x^2).$$

Answer: $\omega_n^2 = 31.0673\frac{EI}{L^4 m}$ (Rad/s)

E.3 A spacecraft is mathematically modeled as a cantilevered beam with bending stiffness EI, distributed mass m, and two discrete masses M_1, M_2. The dynamic system is shown in Fig. E.4.

The assumed mode is

$$\Phi(x) = 1 - \cos\left(\frac{\pi x}{2L}\right).$$

Calculate using Rayleigh's principle the natural frequency ω_n corresponding to the assumed mode $\Phi(x)$.

Answer: $\omega_n = \sqrt{\frac{3.0440EI}{(0..2268mL+0.858M_1+M_2)L}}$

E.4 Investigate and report about the Ritz method [1, 4, 5] to obtain natural frequencies.

References

1. Rao SS (2011) Mechanical vibrations 5th edn. Prentice Hall, Upper Saddle River, Number ISBN 978-0-13-21-2819-3
2. Ludolph G, Potma AP, Legger RJ (1963) Sterkteleer. Number Tweede deel in Leerboek der mechanica. Wolters, Groningen, achttiende druk edition
3. Temple G, Bickley WG (1956) Rayleigh's principle and applications to engineering. Dover Publications, New York
4. Leissa AW (2005) The historical bases of the Rayleigh and Ritz method. J Sound Vib 287:961–978
5. Michlin SG (1962) Variationsmethoden der Mathematische Physik. Akademie-Verlag, Berlin

Appendix F
Random Fatigue Estimation

F.1 Introduction

In this appendix, three methods to predict the fatigue damage and fatigue life of parts of structures exposed to random vibrations are presented. Among many methods, three fatigue damage estimation methods are discussed:

- The narrowband method [1, 2].
- The Dirlik methods, [3].
- The Steinberg's three band method [4].

Examples are provided and problems given.

F.2 Statistical Characteristics of Random Processes

In the frequency domain, the random process X is defined by the one-sided PSD $W_x(f)$, with f the frequency (Hz). The statistical characteristics of a stationary process can be described by the moment of the PSD [5].

The nth spectral moment m_n of the PSD $W_x(f)$ at frequency f is given by

$$m_n = \int_0^\infty f^n W_x(f) df. \tag{F.1}$$

For fatigue prediction, analyses up to m_4 are normally used. The even spectral moments represent the standard deviation of the random process X, and its time derivative $\dot{X}$ are

$$\sigma_x = \sqrt{m_0},\ \sigma_{\dot{x}} = \sqrt{m_2}. \tag{F.2}$$

The spread of the random process or spectral width is estimated using the parameter α_i, which has the general form

J. Wijker, *Miles' Equation in Random Vibrations*, Solid Mechanics and Its Applications 248, https://doi.org/10.1007/978-3-319-73114-8

$$\alpha_i = \frac{m_i}{\sqrt{m_0 m_{2\times i}}}, \tag{F.3}$$

where α_2 is most commonly used and is the negative of the correlation between X and $\ddot{X}$. It takes values from 0 to 1. The higher the value to more narrow is the random process in the frequency domain. It approaches 0 as the random process is a white noise process.

Another frequently used spectral parameter is δ (Vanmarke's bandwidth parameter)

$$\delta = \sqrt{1 - \alpha_1^2}. \tag{F.4}$$

The expected peak occurrence frequency ν_p and the expected positive zero-crossing rate (characteristic frequency) ν_0 are defined as

$$\nu_p = \sqrt{\frac{m_4}{m_2}},\ \nu_0 = \sqrt{\frac{m_2}{m_0}}. \tag{F.5}$$

Example

The response of Problem 11.5, Chap. 11 will be used to calculate the statistical characteristics of the PSD of the bending stress. The PSD of the bending stress $W_{\sigma_b}(f)$ is illustrated in Fig. F.1.

The statistical properties of the PSD function of the bending stress $W_{\sigma_b}(f)$ are presented in Table F.1.

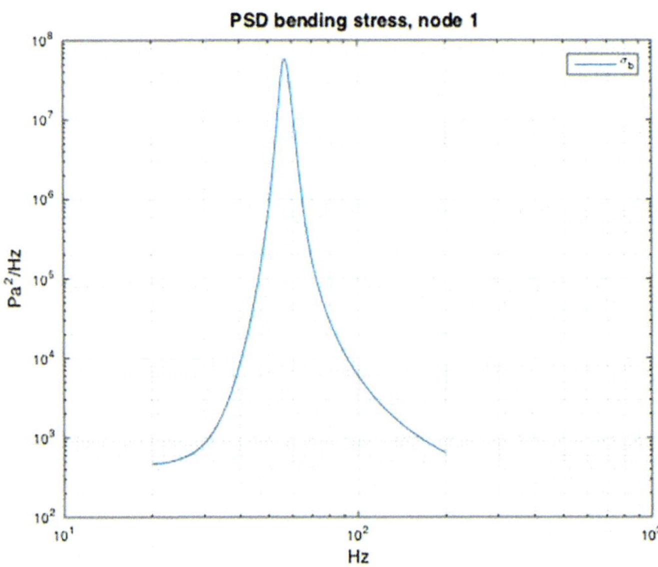

Fig. F.1 PSD bending stress

Table F.1 Statistical properties $W_{\sigma_b}(f)$

$\sqrt{m_0}$ (Pa)	α_2	ν_0 (Hz)	ν_p (Hz)
4.7434×10^7	0.0000	57.3858	58.1468

The PSD response is narrowbanded because the spread factor $\alpha_2 = 0.0000$. The positive zero crossings ν_0 and the averaged number of peaks ν_p are about equal to the fundamental frequency $f_n = 56.2745$ Hz.

The foundation for the frequency domain approximation of a cycle distribution was set by Rice in his famous paper "Mathematical Analysis of random Noise" in [6] where he analytically defined the probability density function of the peaks $p_p(a)$ based on the PSD

$$p_p(a) = \frac{\sqrt{1-\alpha_2^2}}{\sqrt{2\pi}\sigma_x} \exp\left(-\frac{a^2}{2\sigma_x^2(1-\alpha_2^2)}\right) + \frac{\alpha_2 a}{\sigma_x^2} \exp\left(-\frac{a^2}{2\sigma_x^2}\right) \Phi\left(\frac{\alpha_2 a}{\sigma_x\sqrt{1-\alpha_2^2}}\right), \tag{F.6}$$

where a is the peak amplitude. $\Phi(z)$ is the standard normal cumulative distribution function

$$\Phi(z) = \frac{1}{\sqrt{2\pi}} \int_{-\infty}^{z} \exp\left(\frac{-t^2}{2}\right) dt. \tag{F.7}$$

F.3 Fatigue Damage and Fatigue life

The cumulated fatigue damage D is based the Palgren–Miner rule

$$D = \sum_{i=1}^{n} \frac{n_i}{N_i}, \tag{F.8}$$

where N_i indicates number of cycles to failure when the specimen or component is subjected to stress with an amplitude s_i and n_i is the applied number of stress cycles of s_i. Eq. (F.8) indicates that n various stress amplitudes are applied to the component for numbers of cycle. When the damage ratio $D = 1$, the component will fail.

The relation between the occurring stress s and the corresponding allowable number of cycles N can be described by the s-N (Wöhler) curve or Basquin equation

$$s^k N = C, \tag{F.9}$$

where k and C are material constants. The Basquin equation is illustrated in Fig. F.2.

The expected damage $\bar{D}$ per second is based on the Palgren–Miner rule can be estimated by the following expression

Fig. F.2 $s - N$ curve

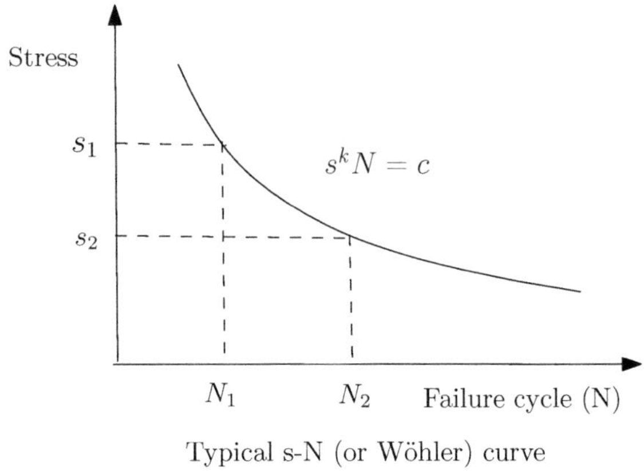

$$\bar{D} = \frac{\nu_p}{C} \int_0^\infty s^k p_a(s) ds. \tag{F.10}$$

The fatigue life estimate is given by

$$T = \frac{1}{\bar{D}}. \tag{F.11}$$

Among many methods in the frequency domain, three methods to estimate the fatigue damage $\bar{D}$ per second will be discussed in this appendix:

- Narrowband method to estimate $\bar{D}_{NB}$ [1, 2].
- Dirlik method to estimate $\bar{D}_{DK}$ [3].
- Steinberg's three stage method to estimate $\bar{D}_{SB}$ [4].

F.4 Narrow-Band Method

For a narrowband random process, it is reasonable to assume that every peak is coincident with a cycle and that the cycle amplitudes are Rayleigh distributed (see Eq. (F.6) with $\alpha_2 = 0$). The narrowband expression was originally presented by Miles [2] and is here defined for stress amplitudes

$$\bar{D}_{NB} = \nu_0 C^{-1} (\sqrt{2m_0})^k \Gamma \left(1 + \frac{k}{2}\right), \tag{F.12}$$

where ν_0 is the expected zero-crossing frequency with positive slope, which is very closed to the peak intensity ν_p for the narrowband process. C and k are material fatigue constants, m_0 is the zero-order spectral moment. For a narrowband process, the parameter $\alpha_2 \approx 0$. The gamma function is defined as

$$\Gamma(x) = \int_0^\infty t^{x-1} \exp(-t)dt. \tag{F.13}$$

F.5 Dirlik Method

The Dirlik method [3, 7] approximates the cycle-amplitude distribution by using a combination of one exponential and two Rayleigh probability densities. It is based on numerical simulations of the time histories for two different groups of spectra. This method is considered as one of the best to estimate fatigue damage. The rain-flow-cycle amplitude probability density estimate is given by

$$p_a(s) = \frac{1}{\sqrt{m_0}} \left[\frac{D_1}{Q} \exp\left(-\frac{Z}{Q}\right) + \frac{D_2 Z}{R^2} \exp\left(-\frac{Z^2}{2R^2}\right) + D_3 Z \exp\left(-\frac{Z^2}{2}\right) \right], \tag{F.14}$$

where Z is the normalized amplitude and x_m is the mean frequency as defined hereafter

$$Z = \frac{s}{\sqrt{m_0}}, \quad x_m = \frac{m_1}{m_0} \sqrt{\frac{m_2}{m_4}}, \tag{F.15}$$

and the parameters D_1, D_2, D_3, R, and Q are defined as

$$\begin{aligned} D_1 &= \frac{2(x_m - \alpha_2^2)}{1 + \alpha_2^2}, \quad D_2 = \frac{1 - \alpha_2 - D_1 + D_1^2}{1 - R}, \\ D_3 &= 1 - D_1 - D_2, \quad R = \frac{\alpha_2 - x_m - D_1^2}{1 - \alpha_2 - D_1 + D_1^2}, \\ Q &= \frac{1.25(\alpha_2 - D_3 - D_2 R)}{D_1}, \end{aligned} \tag{F.16}$$

while α_2 is already defined in Eq. (F.3). The rain-flow damage intensity under the Palgren–Miner rule is calculated by substituting $p_a(s)$ Eq. (F.14) in Eq. (F.10)

$$\bar{D}_{DK} = \frac{\nu_p m_0^{k/2}}{C} \left[D_1 Q^k \Gamma(1 + k) + \sqrt{2}^k \Gamma\left(1 + \frac{k}{2}\right) \left(D_2 |R|^k + D_3\right) \right]. \tag{F.17}$$

The parameters C and k are defined in Eq. (F.9).

F.6 Steinberg's Three Band Method

The assumption of the Steinberg's method is that the probability density function of the random stress response follows a Gaussian distribution, and therefore the expected values of the stress response amplitudes are bound by certain probability levels:

68.27% Chance that the amplitude of the stress cycles will not exceed the peak of one times the r.m.s. of the stress response signal.
27.1% Chance that the amplitude of the stress cycles will not exceed the peak of two times the r.m.s. of the stress response signal.
4.3% That the stress cycles will not exceed the peak of three times the r.m.s. of the stress response signal.

There are no stress cycles occurring with peaks greater than three times the r.m.s. value of the stress.

The expected fatigue damage per second is given by:

$$\bar{D}_{SB} = \frac{\nu_o}{C}\left[0.683\left(\sqrt{m_o}\right)^k + 0.271\left(2\sqrt{m_o}\right)^k + 0.043\left(3\sqrt{m_o}\right)^k\right]. \tag{F.18}$$

Example

The 1σ shear stress $\tau_p = 0.07637$ MPa of Problem 2.10 is used to calculate the fatigue life time of the honeycomb core. The natural frequency $f_n = 75$ Hz. The following (artificial) s-N curve is applied in the fatigue damage calculations

$$s^6 N = 1.9847 \times 10^{35} \text{ cycles Pascal}^6. \tag{F.19}$$

The narrowband method and the Steinberg's method are applied to estimate the fatigue life T of the honeycomb core.

Narrow-band method $\bar{D}_{NB} = f_n c^{-1}(\sqrt{2}\tau_p)^k \Gamma\left(1+\frac{k}{2}\right) = 0.0036$,
$T_{NB} = 1/\bar{D}_{NB} = 278$ s,
Steinberg's method $\bar{D}_{SB} = f_n \tau_p^k c^{-1}\left[0.683 + 0.271 \times 2^k + 0.043 \times 3^k\right] = 0.0037$,
$T_{SB} = 1/\bar{D}_{SB} = 270$ s.

Problems

F.1 This problem is a continuation of Problem 11.5, in particular the FEA part, which will be more or less recapitulated. The natural frequency of the dynamic system illustrated in Fig. 11.14 will be calculated using the finite element analysis (FEA) method. The beam will be modeled with one Bernoulli beam element with the following mass and stiffness matrices [8].

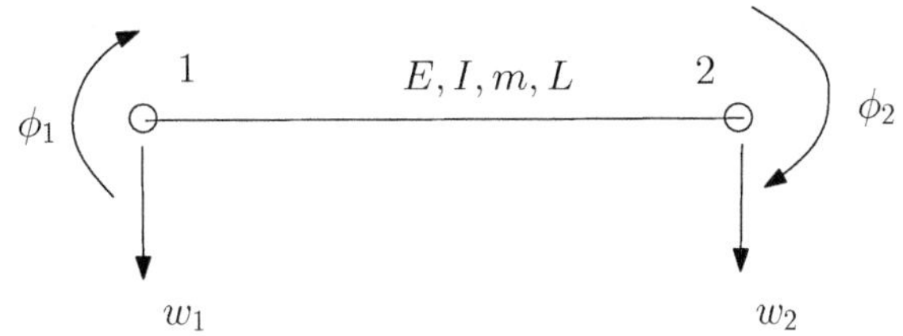

Fig. F.3 Finite element model beam

$$[M]=\frac{mL}{420}\begin{bmatrix}156 & 22L & 54 & -13L\\ 22L & 4L^2 & 13L & -3L^2\\ 54 & 13L & 156 & -22L\\ -13L & -3L^2 & -22L & 4L^2\end{bmatrix}[K]=\frac{EI}{L}\begin{bmatrix}12 & 6L & -12 & 6L\\ 6L & 4L^2 & -6L & 2L^2\\ -12 & -6L & 12 & -6L\\ 6L & 2L^2 & -6L & 4L^2\end{bmatrix} \quad \text{(F.20)}$$

The finite element model (FEM) is illustrated in Fig. F.3.

To introduce the fixation, the large mass approach will be applied (mass $M_l = 10^9$ kg and second moment of mass $I_l = 10^9\,\text{kgm}^2$) in the direction of w_1 and ϕ_1. The updated mass matrix is

$$[M]=\frac{mL}{420}\begin{bmatrix}156+M_l\frac{420}{mL} & 22L & 54 & -13L\\ 22L & 4L^2+I_l\frac{420}{mL} & 13L & -3L^2\\ 54 & 13L & 156+M\frac{420}{mL} & -22L\\ -13L & -3L^2 & -22L & 4L^2\end{bmatrix}, \quad \text{(F.21)}$$

where the mass of the box M is introduced too. To calculate the unitary frequency response functions $H(f)$ of the accelerations at node 1, degrees of freedom (DOF) w_1 and w_2 a constant enforced acceleration $\ddot{U} = M_l$ will be applied to node 1, DOF w_1. That means that the resulting enforced acceleration is equal to one. This is principle the large mass approach. Later on the PSD of the accelerations will be computed.

The natural frequencies f_n and corresponding vibration $[\Phi]$ modes are given hereafter

$$(f_n)=\begin{pmatrix}0.0000\\ 0.0000\\ 56.2745\\ 1503.18\end{pmatrix}, [\Phi]=\begin{bmatrix}0.0000 & -0.0000 & -0.0000 & 0.0000\\ 0.0000 & 0.0000 & -0.0000 & 0.0000\\ 0.0000 & -0.0000 & 2.1820 & 0.3899\\ 0.0000 & 0.0000 & 21.7353 & 340.4000\end{bmatrix} \quad \text{(F.22)}$$

The orthogonal damping matrix $[C]$ can be obtained using Eq. (11.10), with $\zeta = 0.05$.

The PSD of the accelerations of DOFs w_1 and w_2 as well as the bending moment in node 1, ϕ_1 direction are shown Fig. F.4.

The r.m.s. acceleration of the discrete mass M is $w_{2,rms} = 20.9057$ g, and the r.m.s. of the bending moment in node 1 is $M_b = 5.8107\,\text{Nm}$, and the r.m.s. of the bending stress is $\sigma_b = 47.4341$ MPa. Use this stress to calculate the fatigue lifetime with the following methods in conjunction with the s-N curve Eq. (11.44):

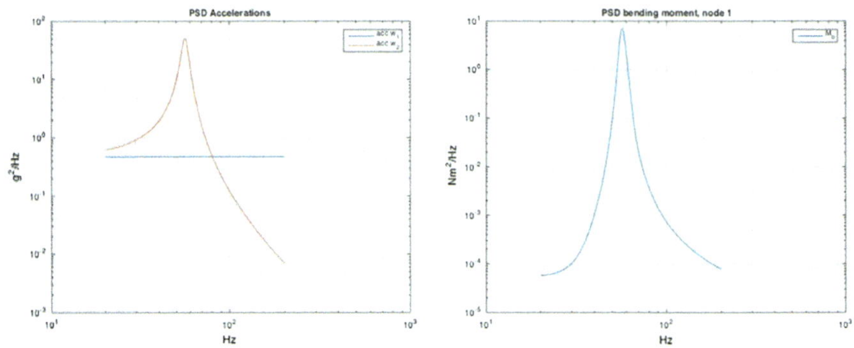

(a) Accelerations (b) Bending moment

Fig. F.4 PSD accelerations and bending moment

- Narrowband method, $T_{NB} = 1/\bar{D}_{NB}$.
- Dirlik method, $T_{DK} = 1/\bar{D}_{DK}$.
- Steinberg method, $T_{SB} = 1/\bar{D}_{SB}$.

Answers: $T_{NB} = 4.0360 \times 10^4$ s, $T_{DK} = 4.0624 \times 10^4$ s, $T_{SB} = 3.9839 \times 10^4$ s.

Repeat the FE analysis with your own favorite FE package and perform again the fatigue life analyses.

References

1. Crandall SH, Mark WD (1973) Random vibration in mechanical systems. Academic Press
2. Miles JW (1954) On structural fatigue under random loading. J Aeronaut Sci 21(11):753–762
3. Dirlik T (1985) Application of computers in fatigue Analysis. PhD thesis, University of Warwick, Coventry, England
4. Steinberg DS (2000) Vibration analysis for electronic equipment, 3rd edn. Wiley, New York. ISBN 0-471-37685-X
5. Mrsnik M, Slavic J, Boltezar M (2013) Frequency-domain methods for a vibration-fatigue-life estimation-application real data. Int J Fatigue 47:8–17. https://doi.org/10.1016/j.ijfatigue.2012.07.005
6. Wax N (1954) Selected papers on noise and stochastic processes. Dover publications. ISBN 0-486-60262
7. Benasciutti D (2004) Fatigue analysis of random loadings. PhD thesis, Universty of Ferrara, Department of Engineering, December 2004
8. Kwon YW, Bang H (2000) The finite element method, using MATLAB 2nd edn. CRC Press, Boca Raton. ISBN 0-8493-0096-7

Appendix G
John Wilder Miles (1920–2008)

G.1 Obituary Notice

John W. Miles, renowned scientist and research professor emeritus of applied mechanics and geophysics at Scripps Institution of Oceanography, UC, San Diego, died October 20, 2008, in Santa Barbara, Calif., following a stroke. He was 87 years old [1]. Miles had been at Scripps since 1964 as a researcher and professor in the Cecil H. and Ida M. Green Institute of Geophysics and Planetary Physics (IGPP) and also served as vice chancellor for academic affairs at UC San Diego from 1980 to 1983.

Miles was well regarded for his pioneering work in theoretical fluid mechanics and in 1957 proposed a wind-wave growth model, a theory he continued to refine until late in his career. His theoretical model is considered as one of the cornerstones for the current generation of numerical wave prediction models, and a major contribution to the field of ocean weather forecasting and wave dynamics. In 2003, researchers at Johns Hopkins University and UC Irvine provided the first experimental measurements supporting important aspects of Miles' theory.

Miles devoted the first 20 years of his research to electrical and aeronautical engineering. When he joined Scripps, he turned his mathematical abilities to geophysical fluid dynamics and made numerous contributions to all aspects of fluid dynamics, including supersonic flow, ocean tides, the stability of currents and water waves and their nonlinear interactions, along with extensive work in the application of mathematical methodology.

'John was a very generous colleague with a great love for research and an immense appetite for work', said Rick Salmon, Scripps professor of oceanography and colleague of Miles.

Miles was born in Cincinnati, Ohio, in 1920, and received his B.S. in 1942, M.S. in 1943, and A.E. and Ph.D. in 1944, all from the California Institute of Technology. During World War II, he worked at the Massachusetts Institute of Technology?s Radiation Laboratory and for Lockheed Aircraft Corporation, followed by an appointment as professor of engineering and geophysics at UCLA from 1945 to 1961 (Fig. G.1).

J. Wijker, *Miles' Equation in Random Vibrations*, Solid Mechanics and Its Applications 248, https://doi.org/10.1007/978-3-319-73114-8

Fig. G.1 John W. Miles 1920–2008

He has received several prestigious awards, including the Timoshenko Medal of the American Society of Mechanical Engineers in 1982, and was designated the Otto Laporte Lecturer by the American Physical Society in 1983. He was elected a member of the National Academy of Sciences in 1979.

Miles' research productivity was legendary, with more than 400 publications over his working 60-year career. He served as associate editor of the Journal of Fluid Mechanics and editor/coeditor of several other scientific journals, covering U.S. and foreign research in fluid mechanics, applied mechanics and mathematical analyses. Within the frame of this book the following publication is of great interest: **'On structural fatigue under random loading',** ***J. Aero. Sci.*** **21, 753-762 (Nov., 1954)**.

He was a longtime resident of La Jolla, Calif., and is survived by three daughters, Patsy Fiske of Santa Barbara, Calif., Ann Leslie Albanese of Solvang, Calif., and Diana Jose of San Diego, Calif. At his request there will be no memorial services. His ashes were scattered at sea.

Reference

1. Reisewitz A (2008) Obituary notice distinguished scientist and professor: John W. Miles. UC San Diego, News Center. http://ucsdnews.ucsd.edu/archive/newsrel/science/10-08JohnWMiles.asp

Index

J. Wijker, *Miles' Equation in Random Vibrations*, Solid Mechanics and Its Applications 248, https://doi.org/10.1007/978-3-319-73114-8

The manufacturer's authorised representative in the EU is Springer Nature Customer Service Centre GmbH, Europaplatz 3, 69115 Heidelberg, Germany. If you have any concerns regarding our products, please contact ProductSafety@springernature.com

Printed and bound by CPI Group (UK) Ltd, Croydon, CR0 4YY
01/10/2026
02225871-0001